Don't Let Technology Crack Your Nest Egg

Don't Let Technology Crack Your Nest Egg

Rethinking Personal Finance for the Digital Age

Ken Kamen

with Dale Burg

SelectBooks, Inc.
New York

This edition published by SelectBooks, Inc.
For information address SelectBooks, Inc., New York, New York.
First Edition
ISBN 978-1-59079-494-4

Library of Congress Cataloging-in-Publication Data

Names: Kamen, Ken, author. | Burg, Dale, editor.
Title: Don't let technology crack your nest egg : rethinking personal finance for the digital age / Ken Kamen ; with Dale Burg.
Description: First edition. | New York : SelectBooks, 2020. | Includes bibliographical references and index. | Summary: ""Financial advisor explores technology's transformation of every aspect of our lives with its potential to both enhance our opportunities and threaten our lifestyle. He recommends the best approaches to protect our 'cyberselves' from manipulation by influencers from ominous sources and maintain control on the digital highway as we spend, save, and invest our assets"-- Provided by publisher.
Identifiers: LCCN 2019034546 | ISBN 9781590794944 (hardcover) | ISBN 9781590794975 (ebook)
Subjects: LCSH: Finance, Personal. | Finance, Personal--Technological innovations. | Computer security.
Classification: LCC HG179 .K356 2020 | DDC 332.024--dc23
LC record available at https://lccn.loc.gov/2019034546

Book design by Janice Benight

Manufactured in the United States of America
10 9 8 7 6 5 4 3 2 1

Dedicated to

Andrea

My best friend, wife, and co-builder of our nest egg

Contents

Preface

DRIVING TO MY OFFICE LAST FALL, I was delayed in a minor traffic backup. Drivers had halted to allow a gaggle of migrating Canada geese to cross the road.

The annual voyages birds make between their breeding and wintering grounds are among nature's greatest miracles. Some geese make roundtrip journeys of 5,000 miles at speeds of 50 mph or more, at altitudes of 30,000 feet, through thin air, and in temperatures that drop to -60° F.

Along with their extraordinary strength and endurance, they are born with the intuitive knowledge to follow a precise flight path that includes way stations for resting and refueling. Flying in V formation, Canada geese know instinctively how to alternate being lead bird and to position their wingtips and synch their flapping in order to catch the preceding bird's updraft and save energy.

Yet they lack the ability to distinguish sky from glass and asphalt from grass. Some die by flying into skyscrapers they mistake for open air, and many more are doomed when crossing roads that to them are just a wider trail. Evolution, which so successfully ensured their species' survival for the 10 million years they've been on Earth, did not prepare them for the perils that have arisen in less than 150 years in a world transformed by technology.

That moment on the road led me to think about the parallels between those geese and ourselves. Evolution readied us, like the geese, to be builders of nests and navigators of the globe. In addition, it gave us the capacity to reach beyond our physical limits—to imagine and build innovations that have made our lives better, and, for the most part, safer.

But just as skyscrapers and highways present menaces to which the geese migrating for survival are vulnerable, unrecognized dangers lurk for us as we hurtle along the digital speedway.

As a financial professional who entered the field when personal computers and the internet were in their infancy, I have watched modern technology reshape our world on a grand scale. I have watched its transformative effect on how we spend our time and money and how it threatens to invade our privacy and take control of our choices. This book is meant to help alert you to the enormity of these changes and suggest and inspire responses and tactics that will help you meet your lifetime goals.

Introduction

The Acceleration of Change and the Decline of Planning

The development of the computer, and the extraordinary evolution of its data processing capabilities, has put incredible power in the hands of billions of people all over the planet in a period so brief that it barely registers on the timeline of human development.[1]

Technological advances have indisputably made our lives easier, more comfortable, more delightful, and in general (when in appropriate hands) less perilous. We can communicate with one another across great distances and operate more efficiently in every facet of our lives.

In my own case, acquiring one of the first generations of personal computers—machines small enough to fit on a desktop—had a very direct and extremely consequential impact.

In the fall of 1980, I was a college senior. Like many of my fellow students, I hoped the courses we took in quantitative analysis, portfolio construction and computer programming would enable me to secure a perch on the lowest (unpaid) rung of a financial firm's corporate ladder. When I scored one, I was confident that I was on my way to the big time as a financial analyst.

Initially, I worked in the back office reviewing spreadsheets and preparing reports, but very soon the front office operations caught my eye. Brokers there spent their day prospecting for, and talking with, clients. The brokers were the people who generated the mountain of papers I was scaling, and—more important—they were the ones who were driving the firm's business. So much for being an analyst: I decided the front office was the place for me.

So, while completing my internship, I studied for the Series 7 exam to become a licensed stockbroker. I passed it, and upon graduating, I was working as a fully-commission-based Registered Representative. Translation: No paycheck unless I built a clientele. In 1981, that meant cold calling. I compiled lists of prospects at home and in the library on weekends, and every weekday I dialed about 300 phone numbers.

Most people just hung up on me immediately. Some were polite enough to allow me to deliver my pitch before they responded with some version of "Thanks, but no thanks." Ten to twenty a day might say, "Call when you have a stock recommendation worth sharing." But when I did, most of these "warm" leads cooled, and many didn't seem to remember my previous call.

Every broker in the office at the time went through this ridiculous dance. Back then, building a business had way more to do with being undaunted by rejection than with presenting promising ideas. Often I went home thinking I had chosen the wrong career.

But one night, I had an idea that turned out to be life-changing: that I might have better results if I followed up each call with a personalized communication, something that a prospect could actually have in hand. No other broker in the office was doing this because none of us had the time to type an individual letter to everyone we contacted.

Being fresh out of school, though, I had a powerful advantage over my colleagues: I had learned to use a computer. I realized it was the way to get my letters written, and I decided to go for broke. To buy one of the newly introduced models, a Kaypro, along with a pinwheel printer and supplies, I took out a loan for $3,800—about $11,800 in 2019 dollars—and agreed to pay it off at the rate of $120 a month.

No early computer could manage a contact list with the flexibility and ease of today's models, but it could take a list of names and addresses and merge them one by one into a form letter that would look as if had been typed individually. However, the dot-matrix printers most common at the time accepted only paper that came in rolls with holes down each side, so the documents they printed looked mass-produced. That's why I had splurged on the pinwheel printer. It took conventional sheets of paper, but it was ear-splittingly noisy.

Working out of a cubicle in those days—I had yet to merit either an assistant *or* an office—I had to buy a soundproof box to put the printer in. It was still making enough of a racket to be distracting, so I went into the office at night with my then-girlfriend Andrea (still my girlfriend but now also my wife and the mother of our two daughters). We'd work for hours, with me inputting the day's list of names into the computer while she folded and stuffed envelopes.

The payoff was tremendous. Because I had taken the time to send them a letter, as many as four out of five people would engage with me on the second call—a much higher percentage than any of my colleagues experienced—and I could continue the conversation rather than reboot it. That Kaypro computer is still in our basement (despite Andrea's attempts to get me to toss it); I keep it as a reminder of how investing in myself helped me turn lukewarm leads into legitimate prospects. That's what cutting-edge technology did for me.

Change at an Unprecedented Pace

Of course, I could not have anticipated the magnitude of the changes that personal computers, together with the subsequent introduction of the internet, would bring about. I could never have imagined that before long, more than half the population of the world would possess a relatively affordable wireless device, small enough to slide into a shirt pocket, that would almost magically allow you to handle multiple tasks:

- Communicating with close to half the world's population, anywhere, in real time, without any technical expertise and in their own languages
- Researching information from billions of sources, on any subject, in seconds
- Transmitting and watching events worldwide as they happen
- Accessing unlimited amounts of music and other entertainment on a whim
- Shopping 24/7 in locations around the world
- Launching a business with virtually no overhead

Alvin Toffler was one of the few who foresaw what lay ahead. In his 1970 book *Future Shock,* he suggested that the accelerating pace of technology would precipitate a "dizzying disorientation brought on by the premature arrival of the future."

"Change is avalanching upon our heads," he presciently wrote, and "most people are grotesquely unprepared to cope with it."

When we speak of "technology" today, generally we're referring to electronic devices, especially computers and smartphones, and the internet. But the word, from ancient Greek, has a broader meaning. It derives from "techne" meaning an art or skill, and the entire word refers to the practical application of knowledge, especially in a particular area. In its larger sense, "technology" can refer to anything considered a tool.

The use of tools is one of the key abilities that set us apart from other species. The development of tools made it possible for Homo sapiens to evolve. Dramatic technological advances have changed the course of human history many times, but historically they occurred only over huge spans of time.

About 2.4 million years ago, hominids began making tools by splitting pebbles. As of 1.6 million years ago, they had learned how to create a sharp edge on stones; that let them cut through hide so as to get meat off bones and extract marrow. These new sources of nutrition, scientists speculate, allowed them to move to habitats where they didn't have to rely on vegetation to survive.

About one million years ago, man learned to create "controlled fire" to cook food, which released additional nutrients and helped prevent disease. It also meant they could spend less time foraging and were able to clear land, which laid the "groundwork"—literally—for agricultural development.[2]

The next monumental development, the invention of spears for hunting, took another 600,000 years, and the one that followed—the invention of boats—came only 540,000 years later.

Potters probably invented the wheel around 5000 B.C., but combining it with the axle to make vehicles for transportation purposes was the eureka move. To develop tools that could make the ends of the axles and holes in the wheel nearly perfectly smooth and round took another couple of thousand years.[3]

These developments helped humans grow healthier, stronger bodies and—even more significantly—bigger brains, which allowed them to develop both intellectual curiosity and the ability to venture to distant places.

The intervals between big ideas and improved iterations of those ideas began to shorten. In weaponry, for example, just nine centuries passed between the use of gunpowder in warfare (about 1000 A.D.) and the development of the atom bomb in 1945.

Benjamin Franklin flew a kite into a thunder cloud to demonstrate the nature of electricity in 1752. In fewer than 150 years, that insight was turned to practical use. In 1891, London's Savoy Theatre became the world's first public building to be lit entirely by electricity.

Functional, Mechanical, and Cognitive Tools

The earliest tools helped our ancestors meet their functional needs for hunting, farming, building, and other activities. Eventually, more sophisticated mechanical tools were developed that helped them complete tasks way beyond their physical capacity.

In the 1990s, author Sharon Derry established a new category she called "cognitive tools" that she defined as mental and computational devices that support, guide, and extend the thinking process of their users. She was referring to communications software such as teleconferencing programs, databases, and spreadsheets.

There were earlier precursors, I believe. I would count the development of language among them. Language gave us the capability to describe abstract concepts and to think on a higher plane than a pragmatic one. The development of writing, and then paper, allowed ideas to be passed down from generation to generation without the limitations of memory; and the subsequent invention of the printing press allowed wide dissemination of those ideas. Now, of course, much of humanity's collective knowledge is instantly available to anyone with a smartphone.

Before computers, cognitive tools—like other tools—were introduced over relatively long periods of time, but that has not been the case in the last couple of centuries.[4] In 1832, British inventor and mathematics professor Charles Babbage designed a forerunner to the computer he called the "first difference engine," a hand-operated mechanical calculator. Fewer than 50 years later, and in a span of 36 months, inventor Herman Hollerith devised a punch card system that was used to calculate the 1880 census. (He also founded the company that became IBM.) In 1936, English mathematician Alan Turing presented a theoretical machine capable of solving any problem that could be described by simple instructions encoded on a paper tape. Within a decade came the first operational computers that were huge, tremendously expensive machines.

The real game-changer was the invention in the 1970s of a microprocessor, the central processing unit on a single integrated circuit chip containing millions of very small components that lets the computer do its work. The pace and impact of technological change then began to race forward.

By 1977, three "personal computers" with mass appeal were introduced: Apple, Inc.'s Apple II, Commodore Business Machines' Personal Electronic Transactor and Tandy Radio Shack's TRS-80. (My revered Kaypro appeared five years later.) The precursor to the internet was introduced in the 1980s, and a decade later, the worldwide web was invented. That allowed everyone with a computer and internet connection to connect with one another and to access an almost incomprehensible amount of information, all in real time.

At the same time, something remarkable was happening to telephones. In 1992, IBM introduced a prototype mobile phone that incorporated PDA (personal digital assistant) features. Two years later, Bell South brought out the Simon Personal Communicator, the first real smartphone. It could make and receive calls and also send facsimiles.

The iPhone, Android, and Kindle arrived in 2007, which Pulitzer-Prize-winning political commentator Tom Friedman identified as the exact year when, as Netscape cofounder Marc Andreessen put it, software began "eating the world."

The iPhone and its subsequent imitators put the internet in everyone's pocket. Plus, the single palm-sized item soon incorporated the capabilities of a remarkable number of items: a clock, alarm clock, and watch; a calendar, date book, and address book; maps and a compass; a flashlight and level; a camera and video camera; postcards and photo albums; a scanner; an iPod; every sort of book, newspaper, and magazine; a notebook and a pen; a computer; software creation and distribution (via app); and, of course (though initially, sometimes unreliably) a land line.

As humorist Dave Barry had observed about his smartphone, "With this amazing device, I can send and receive e-mail and text messages, surf the internet, pay my bills, book flights, play games, take pictures, listen to music, watch TV shows—in short almost anything except reliably make or receive telephone calls. For some reason, cellular telephones lack that capability. It's as if they made a washing machine that mowed your lawn and made daiquiris, but if you put your actual clothes into it, they burst into flames."[5]

Their occasional issues notwithstanding, cell phones were quickly and widely embraced, with myriad consequences. Some could have been expected: Mobile advertising sales quickly surpassed desktop advertising sales. Others might have been a surprise: Starting in 2007, gum sales in supermarkets plummeted. While bored customers standing on line used to make impulse buys of packs of gum, now, with eyes fixated on their phones, they're oblivious to such distractions.[6]

In April 2019, a huge percentage of the world's population had digital access, and the numbers were growing rapidly. (The year-over-year increase from April 2018 in each category is reflected in parentheses.)

- Total population: 7.67 billion (+1.196%)
- Unique mobile users: 5,110 billion, or 66% penetration (+2.6%)
- Internet users: 4.437 billion, or 58% penetration (+896%)
- Active social media users: 3.499 billion, or 45% penetration (+6.1%)
- Mobile social media users: 3.429 billion, or 45% penetration (+11%)[7]

Advances in technology have transformed the way we hear news, conduct business, receive health care, solicit advice, buy and sell, socialize, and more. Thanks to the ubiquity of its presence, the intimacy of our connection to it, and above all, the pace at which it is evolving, technology is certain to have long-term, uniquely dramatic, and enormous effects on how our species learns, thinks, behaves, and evolves.

Observing that the number of transistors per square inch on integrated circuits had doubled every year since their invention,

Intel founder Gordon Moore in 1965 predicted that integrated circuits would have twice the computer capability of the previous generation for the same cost of production.[8]

True, the pace has slowed—the number has now doubled only every 18 months, which is how "Moore's law" is currently defined—and some believe there will be a limit to this law. But change itself will continue, and at unprecedented speed, exactly as Toffler warned decades ago.

In *Thank You For Being Late*, Tom Friedman explained that it now takes us only 10 to 15 years to get used to the sort of technological changes that we used to take a couple of generations to absorb. But that's not fast enough, considering that technology becomes obsolete every five to seven years. In the two and a half years he spent researching his 2017 book, he had to interview the main technologist at least twice because things changed so quickly. (With this book, I've had some updating challenges myself!)

The Decline of Planning

My world, like virtually every other human enterprise, has been operating at an increasingly accelerated pace. It's almost incredible to recall that years ago, if the New York Stock Exchange had a heavy trading day, the market would open late the next day to allow time to deal with the paperwork. And if it were a *really* heavy trading day, the Exchange might even close for a day and a half to give the clerks time to catch up on all the orders.

That practice was long gone when I started out as a broker in the early 80s, although things still were relatively slow and deliberate. Today, there is almost no lapse between the time when an event occurs and when news of it is broadcast; news comes to us 24/7 and flux is constant. Buying and selling is continuous, global news and market updates are streamed in, messages are sent and retracted, appointments are scheduled and canceled. In response, we have collectively gotten *into* the habit of acting impulsively and *out of* the habit of thinking ahead. That's a huge and unfortunate development.

I noted earlier that the use of tools is one of the key traits that sets us apart from other species. But it may not be the most important one. "What best distinguishes our species is an ability that scientists are just beginning to appreciate: We contemplate the future. Our singular foresight created civilization and sustains society," write psychologists Martin E.P. Seligman and John Tierney.[9]

Life was unsustainable without planning. Humans had to learn to plant in the spring in order to harvest in the fall and then to stockpile food for seasons when it wasn't plentiful. Because of the time-consuming labor involved to produce clothing, tools, and shelter, needs had to be anticipated long in advance.

The ability to plan allowed us to consider increasingly distant goals, to make journeys in order to explore unknown places, to envision structures that might take years or decades to complete, to create rules of law and governance that held for centuries.

In fact, Seligman and Tierney propose that our ability to look ahead suggests we might be called Homo prospectus rather than Homo sapiens. "Looking into the future, consciously and unconsciously, is a central function of our large brain."[10]

Yet although we now have unprecedented capacity to prepare for and shape our future—there have never been so many resources, so many advisors and consultants ready to help us—I am seeing signs of an unexpected and oddly ironic phenomenon: the decline of planning. Like change itself, it also is happening at an accelerated rate.

People currently in or anticipating retirement were raised by a generation that had the safety net of social security and pensions to look forward to when they left the work force. And thanks to the prolonged post-war economic boom, they were able to build a nest egg fairly easily. But having been raised in families who lived through the Great Depression, they learned lessons and habits of prudence and planning they passed on to their children.

However, those old lessons and habits faded from the collective psyche by the time today's young adults emerged. Today, you don't buy things on a layaway plan; you use credit cards. You don't wait for something until you've paid for it. You get it first, and then you pay for it—over time if you choose to (and many do), incurring interest charges that keep you in debt, possibly for years.

A few decades ago, a college student had to anticipate running short of cash well before it happened because solving the problem took time. You'd likely start by writing home with a request for more (making a collect long-distance call, over a shared dorm phone, was a rare indulgence), wait for a check to arrive, deposit the check in the local bank, and then wait for it to clear before the money was available. Before the introduction of cash machines in 1974, that meant going to banks that were open only on weekdays between 9 and 3. Estimating the pace at which various parts of the process would proceed was a critical part of learning how to plan.

As of 20 years ago, students could text a parent for cash, wait briefly for the electric transfer, and get the cash from any nearby bank machine almost immediately. With Venmo, it's even faster: funds can be transferred from phone to phone.

Technological advances have helped create a culture of instant gratification. Books and surfboards, ponchos and keyboard covers, poison remedies and Valentines—almost anything you can think of can be ordered virtually as soon as the need or whim demands and in just a day or two, sometimes within hours, delivered directly to your urban or suburban home. Perhaps in the near future with the aid of drones, things will arrive just as rapidly to remote locations. 3-D printers may become avatars for fairy godmothers. Wish for something and presto! It will appear in your living room.

We have gotten out of the habit of planning ahead. Yet it is more critical than ever. That's because—again, thanks to technology that has made us healthier due to advances in agriculture, better drugs, and improvements in health care—our life expectancy has been extended so significantly we may be underprepared to meet our financial needs for such a long future.

Meanwhile, two important support systems, our Social Security and pensions, are in danger of being reduced or disappearing entirely.

Created in 1934, Social Security established the concept of retirement as a general entitlement and set retirement age at 65. In that Depression era, it was meant to ease out older workers and free up jobs for young male heads of households. Men were the primary wage earners, and their life expectancy then was just 59.3 years, so Social Security didn't have a transformative effect.[11]

During the following decade, the booming post-war economy led businesses to offer perks—notably pensions—to attract employees. Together with Social Security and personal savings, they made a comfortable retirement very achievable, especially because it was still likely to be a short one. People born in 1950 (all races, both sexes) had a life expectancy of only 68.2 years.

The numbers have changed remarkably. By 2016, the Social Security Administration calculated that a 65-year-old man (born in 1951) could expect to live another 20 years (not just the additional 3.2 or so predicted at his birth), and a 65-year-old woman was likely to live another 21. In fact, one-fifth of men and one-third of women in their 60s can expect to live until age 90. This is an exhilarating and also sobering development.

The population of the United States is aging, and very likely there soon will be fewer workers paying into the system than there are people of retirement age. The trustees anticipate that within a decade or two the program may have to tap its trust funds to continue to keep benefit payments at their current level, and a serious shortfall within a couple of decades seems likely.

In addition, traditional defined benefit plans, where employers made the total contribution, are disappearing. Like other union members who are now retired, teachers enjoy a comfortable income thanks to their pensions. But that will not be so in the future. A 2019 study reports that 81% of teachers who start working at age 25 won't qualify for adequate retirement benefits under a typical plan.[12]

In industries where there is a strong union presence (such as airlines and automobile manufacturing), some employers still offer traditional defined benefit pensions, but they are definitely on the decline. The percentagc of workers in the private sector whose only retirement account was a defined benefit pension plan was near 60% in the 1980s, but by 2018 that number had declined to 4%. Today about 14% of companies offer a combination of a defined benefit pension plan and a defined contribution plan (like a 401K), which allows you to invest money that won't be taxed until it is withdrawn. In some cases, the employer will match your investment amount.[13]

Future retirees must find a way to pay their current bills during a typical work life of 40 or so years and at the same time be capable and disciplined enough to sock aside an amount large enough to maintain their lifestyle during a retirement period of 20 years . . . or more.[14]

The more robust your lifestyle has been, the larger the gap between Social Security and other income sources and the amount you actually need to fund your lifestyle. Many Americans have become accustomed to living very well. There are more things to buy than ever before, and buying on credit is an alluring temptation. By the end of 2018, Americans had racked up a record $870 billion in credit card debt (up 45% over the last five years) and paid $110 billion in credit card interest and fees (up 13% from the $98 billion in interest paid in 2017). By February 2019 the total American consumer debt exceeded $4 trillion for the first time.

That the economy has been behaving in unprecedented and unanticipated ways has just compounded the problem of long-term planning. And it is a task for which people are increasingly ill-prepared. Financial illiteracy has reached epidemic proportions. Having learned to rely on electronic calculators, fewer people are capable of doing simple math even for estimating purposes, such as quickly glancing at a bill to see if the total is approximately correct or quickly calculating a waiter's tip. Social media ramps up the desire to buy things and experiences, and the virtualization of transactions makes people less aware of what they are spending and saving. Schools have not risen to the challenge of educating students about financial matters and money management.

And although technology has provided us with unparalleled access to a universe of data at a keystroke, it doesn't always make us smarter or more competent. In fact, it may have a contrary effect by creating a false sense of security and the conviction that you don't need to think prospectively.

As you might expect, and as I see often in my work, people tend to put off making significant planning decisions until a motivating incident comes into their field of vision. The death of a peer may give them a push to finally buy life insurance and/or make a will. A sudden market dip may spur them to flee from the stock market.

But decisions made reactively are almost always not the best ones, and the costs may be high both in long-term consequences and in dollars. If you have procrastinated over buying insurance, for example, you may be obliged to pay a higher rate or even (depending on your health and/or age) become ineligible for coverage. If you sell your stock on a short-term dip, rebuying your portfolio after the market rebounds may cost you dearly.

What's more, the very existence of so much computer-generated information about planning issues can result in more confusion than solutions. The internet is built for speed and simplicity and binary thinking. The human brain is capable of more nuance. It can accommodate multiple contrary ideas.

We rely on "external brain" devices—the smartphone or tablet we carry in our pocket or handbag—to keep track of phone numbers and appointments and shopping lists, to search on the internet for facts we've forgotten or never knew, to remember to send a birthday card or turn off the lights we've left on at home.

It seems as if all this convenience would buy us some time here and there for reflection and contemplation, but that doesn't appear to be what's happening. Instead we increasingly turn to devices to divert us with games, Instagram scrolls, YouTube videos, and other distractions.

In fact, the idea of being left with one's own thoughts has become, for some people, an unbearable prospect. Astonishingly, in a small study in which people were allowed to choose between sitting with no distractions for between 6 and 16 minutes or getting a mild electric shock, many chose the shock.[15]

Taking the expression "down time" literally—by putting down the phone and the iPad and the other devices—gives you an opportunity to let your mind roam and do what it is that makes us human.

The brain has two different attention systems, according to some researchers, the external network that is focused on the outside world and the internal or default mode network (DMN), which is active during day dreaming and mind-wandering as well as when you're thinking about yourself, your past, and the future. The networks can't operate simultaneously.[16] When research subjects were asked to think of "nothing," neuroimages showed the DMN remained active. Significantly, this network is linked to the part of the brain connected with "executive" function, which does things like manage time and pay attention. The link suggests that daydreaming might help us think both creatively and productively.[17]

In fact, many groundbreaking achievements were the result of idle observations. Watching an apple fall from a tree gave Sir Isaac Newton the idea that ultimately led him to define the law of gravity. When burrs snagged to his coat and his dog's fur as they walked, George de Mestral, a Swiss engineer, examined them under a microscope. Observing that the plant was made up of tiny threads with hook-like ends that caught onto fabric and fur, he went on to invent Velcro®, which made him a millionaire.

You may not go down in history or hit the jackpot when your own mind wanders, but instead of turning to fidget toys, ubiquitous screens and other distractions, letting your thoughts go where they will during a boring car ride or while sitting in a doctor's waiting room may help you resolve a problem or have an insight that may make your future more secure.

I work with my clients to help them with prospective thinking, an area in which humans—at least up to now—do better than computers. "Humans learn to achieve many types of goals in a huge variety of environments. We don't yet know how to endow computers with the kind of common sense understanding of the world that underpins human general intelligence," says Murray Shanahan, Professor of Cognitive Robotics for the Department of Computing at Imperial College in London.[18]

The combination of common sense understanding plus imagination gets results that can't be produced by devices, which thus far, anyway, allow us only to input things and get an output.

That's why I am dismayed about the decline of planning exhibited by the tendency of people to download the financial facts of their lives to their devices without scrutinizing those facts. When working with new clients, I always start by asking some basic questions about their finances. Not long ago, people would be able to give me the answers off the top of their heads. Today, they have to pull up their screens.

What's more, despite the ease of monitoring financial information, they are not taking responsibility for oversight. This is like the inverse of compounding interest. In compounding, a little bit grows exponentially into a larger accumulation of money. When you fail to take responsibility for your finances, a small amount of inattention grows into an ever-larger knowledge gap, so you're less and less capable of managing them.

You get a false sense of security believing that just googling a topic can get you all the information you need precisely when you need it. But the unintended consequence of living your life as if everything is available on demand is poor decision-making.

To make important choices you need undistracted time for contemplation, time to call up past experiences and roam through data points, and time to put together the bits and pieces stored in your brain. The process can be ongoing, and your brain may return to it in all sorts of places and odd times: in bed before you drift off to sleep, during your morning shower, at the beach, while commuting or waiting in line when your phone has run out of charge—a turn of events that may initially dismay you but prove to be fruitful!

An hour or two of reading before making a decision isn't really planning, particularly if you're getting your information online. The internet makes things buzzy and shiny, and it's fast-paced. It's about novelty and speed, about results rather than process.

Online planning and investing advice based on your answers to a list of questions cannot possibly be reflective of all the considerations that should guide your actions.

What's more, point-and-click decision-making isn't the same as making your own deliberate and unique choices. The very exercise of making your decisions helps you commit to seeing them through.

Long-term planning is the key to long-term security. It's quite different from short-term, *tactical* planning. That's a matter of figuring out how to go from Point A to Point B *when you know where point B is*. Long-term planning is *strategic*: You don't really know what you're planning for, but you have to take some action.

One of the most astute comments I ever heard on this subject came from President Dwight Eisenhower: "Plans are useless, but planning is essential." The former general learned that lesson on the battlefield. Nowhere in the real world do things happen precisely as anticipated, but the experience of considering many scenarios and multiple approaches prepares you to handle whatever comes up.

In my work, I see the consequences of failing to plan strategically all the time. As I tell my clients, at age 30 and 40 you don't typically spend time imagining yourself as a 60- or 70-year-old struggling with unexpected medical bills, having to provide help for your adult children or very senior parents, unable to afford the leisurely retirement you anticipated. If you could see that those future fiscal problems are rooted in your current behavior, you would be much more motivated to change it.

It's never too early to start planning—to anticipate the potholes and land mines, the detours and retreats; to separate the positive from the negative, the helpful from the detrimental; to be open to opportunities and challenges while avoiding threats to long-term security—in order to reach your goal of keeping your nest egg from harm and protecting your lifestyle for a lifetime.

Part One

TRANSFORMATIONS

1

Losing Focus:
Distracted by the Small Screen, Ignoring the Big Picture

What if the cost of machines that think is people who don't?

—George Dyson, technology historian

I WAS TALKING TO A YOUNG COUPLE whose parents had suggested they come to me to get an early start on long-term financial planning. Naturally, one of the topics that came up right away was how they could start to build their future nest egg. The two of them couldn't squeeze a lot of savings out of their current income, so they had both decided to sign up for overtime work at their respective jobs for this purpose. But the husband admitted that he had blown a hole in the plan several months earlier.

"I'm a really devoted video gamer," he said, with a sheepish shrug. "And a new version of the game I really like was released a few months ago. I got absolutely hooked on it. I couldn't wait to get home at night to start playing."

"—And for a month and a half, he didn't put in any overtime," his wife interrupted.

"Really bad move," he said, shaking his head in dismay at his own behavior. "I can't believe I did it."

He's far from an isolated case. A research paper released by the National Bureau of Economic Research found that men ages 31 to 55 were working about 163 fewer hours a year in 2015 than their cohorts did in 2000. Men who were from 21 to 30 years old were working 203 fewer hours a year. What accounted for the difference? Time spent playing video games.[19]

The authors of the study blamed the onset of this phenomenon on the Great Recession that occurred in the early 2000s. Left with more leisure time because they were out of work, men in these age groups—critical years for career building and amassing wealth—spent it on games. (Though in the same period, for the same reasons, women and men in other age groups also had more leisure time, they pursued other activities.)

The authors of the study speculated these young people were playing games rather than devoting more effort to looking for work because they assumed they'd find employment eventually, when they wanted to. Also, the game-playing satisfied them. Apparently, it gives players purpose (improving your skill, performing with peers, impressing the community) and a routine—the very same sources of satisfaction you get from traditional work.

Many devote to it as much time as if it were indeed work. According to The Entertainment Software Association research, the average dedicated gamer is 35 years old and has been playing for 13 years. He spends 6.5 hours a week playing with others online and 5 hours playing with others in person, a cumulative total of nearly 7,800 hours.[20]

Then consider this: You can earn a degree from many colleges by completing 120 credits, usually accomplished by taking 40 three-credit courses over four years. A three-credit course typically requires about 135 hours of your time—45 in the class plus 90 hours of homework. So to complete coursework for a degree takes about 4,700 hours. By devoting even a portion of game-playing activity to earning a degree (or any other specific goal), imagine what a dedicated gamer might accomplish.

But gamers aren't the only ones allowing a multitude of images and messages to suck up vast amounts of time and bandwidth. Captivated by one or another of the categories vying for our attention—entertainment, news flashes, text messages, emails, voice mails, tweets, Facebook postings, Instagrams, and Snapchat—we spend hours in front of screens. Watching is so compelling that some of us actually become addicted to it.

"Technology Addiction" Is Now Officially a Disease

When people are online, their brains show the same reward pathways as brain scans of people who have substance abuse disorders. And like other addictions, internet addiction begins with occasional use, which becomes daily use, and finally escalates to usage that has consequences that might include insomnia, dysfunctional relationships, and school or work absences. As with other addictions, tolerance grows. For example, an addict who's a gamer devotes more and more time to playing a particular game.[21]

But it's not the number of hours you spend on the internet that determines if you're an addict or even specifically what you're doing. Whether you're playing a game, constantly checking Facebook and Instagram, watching YouTube on loop, or reading Reddit nonstop, what matters is that your use may significantly and adversely affect your relationships, health, and career prospects.

In May 2019 the World Health Organization (WHO) for the first time included "gaming" in the category of "Disorders due to addictive behavior" in its 11th International Classification of Disease. "Gaming addiction" and related problems collectively referred to as IADs (Internet Addiction Disorders) have become so common a topic that a new musical in New York in 2019 featured "web addicts" singing in an a cappella chorus against "the digital Goliath." And an industry has sprung up to deal with the problem.

Some psychiatrists now specialize in internet addiction, and there are multiple other treatment choices, from short-term (a Digital Detox $299 overnight hotel package in a tech-free room with a 10% discount for every night you abstain) to very intensive treatments (retreats, wilderness-therapy camps, and in-hospital programs). One residential facility that treats gaming addiction on the West Coast was recently charging nearly $30,000 for seven weeks of care. Its waiting list was months long.

For most of us, of course, excessive focus on screens is likely to affect our pocketbooks in less dramatic but still troubling and significant ways. Even if your usage doesn't rise to the level of addiction, it can negatively impact your relationships, health, and work by impairing your social skills and lowering your productivity. Cumulatively, the consequences can compromise your future.

Most of us don't fully realize the extent of the resources businesses have directed toward finding ways to get us to spend our time in a way that benefits them. One of the people who sounds the alarm in this regard is Tristan Harris, who was initially hired by Facebook as a "design ethicist" to achieve a lofty goal: helping the company figure out how to develop products that were *not* manipulative. But he eventually decided it was an impossible task, and he quit. Even the news feed on Facebook, he has said, is not neutral: It is designed to get you to scroll.

"We're swimming in a digital environment that is . . . handcrafted by a handful of companies with deliberate goals to capture human attention. And now you have two billion people, 25% of the world's population, whose thoughts are shaped by this handful of . . . engineers and designers in California," Harris said on NPR.[22] (Harris, incidentally, helped to popularize the idea of turning off the color on your smartphone and using grey scale to make the screen less stimulating and less tempting. Color and contrast make you more vulnerable to what is called "attentional recruitment." Lifehacker explains how to do it.[23])

As you might expect, the effect on young people of what has been termed "the device culture" concerns many, and by now that even includes the people who were instrumental in creating it. At least one executive who helped develop the iPod and iPhone has ruefully noted how addictive devices can be and mentioned the "unintended consequences" of these devices; another has expressed his "guilt" at building the social network.

The New York Times reported in 2018 that Tim Cook, the CEO of Apple, would not let his nephew join social networks. Bill Gates banned cellphones until his children were teenagers, and Melinda Gates wrote that she wished they had waited even longer. Steve Jobs would not let his young children near iPads.[24]

Looking back, it's not surprising. The goal of Facebook was clear in the beginning: how to consume as much of your time and conscious attention as possible?"[25] Device manufacturers have been remarkably effective at doing that. A 2018 Deloitte Global Mobile Consumer Survey revealed the following:

- Smartphones easily remained the most favored mobile device among US consumers and reached 85% penetration overall.
- Americans are viewing their smartphones more often than ever before, on average 52 times per day.
- More than a third of adults report using their smartphones "very/fairly often" for business purposes outside normal working hours.

And smartphones are only one of many electronic temptations. Cable providers, broadcasters, independent channels, and internet websites operate like schoolyard drug dealers, offering entry-level diversions that may turn you into an addict with a pharmacopoeia of distractions. Just as with drugs, it's easiest to hook younger people, but frustrated adult workers who aren't fulfilled by their jobs are vulnerable as well.

There are hundreds of videos to watch and games to play that satisfy the most specific and obscure of interests; "time-shifted" entertainment that lets you binge-watch on demand, even automatically loading the next segment to keep you glued to your seat—what has been referred to as "hijacking our brains." And the mental junk food of distractions I think of as "shiny stuff" are largely sophomoric, prurient, and/or pointless. (I always think of G.K. Chesterton's remark—"Journalism largely consists of saying 'Lord Jones is Dead' to people who never knew that Lord Jones was alive"—when I'm even remotely tempted to click on "This is What [*Person I've Never Heard of*] Looked Like Then, and You Won't Believe What [*He or She*] Looks Like Today." But the Lord Jones of Chesterton's day probably was more worthy of note than many contemporary "celebrities.")

Sitting outdoors last summer, enjoying a cool drink under a star-filled sky, I had pulled out my cell phone and spent $3.99 on a Celestial Map to pick out the constellations when I suddenly thought, "This is nuts. I'm in my own backyard. I can look at the sky for free. I can just relax. Why am I reaching for electronic entertainment and paying for it, to boot?" Because I'm just as susceptible as the next guy. We don't know how to just sit and do nothing—which actually can be very productive.

Research: Majority of Time Spent on Nonproductive Activity

The providers of what is collectively (and, one might add, optimistically) called "content" ("distractions" would be more accurate) are companies who sell advertising. The more eyeballs they can attract and the longer they can keep them in place, the more they can charge. Meanwhile, they are taking away your most precious resource—time.

Of course, thanks to our increased life spans, we have more time than ever. As I have pointed out, many of the people who reach the age of 65 today will live into their 90s, yet the typical span of work lives, from early 20s to late 60s, is still slightly longer than 40 years. In that relatively limited time, the challenge is not only to cover your living expenses but also to build a nest egg to fund an additional 30 or so years of retirement. This means you should be spending your time more productively than ever during your working decades; but increasingly, that's not the case.

In 2015 Common Sense Media did a research study that found that 31% of teens spent four to eight hours daily with screen media, and 26% spent more than eight. (Screen media include TV, videos, social media, mobile/computer games, and video games.) In some cases, they were engaging in more than one activity at a time.[26] According to a Pew Research study, by 2018 even the teens were worrying that they spent too much time online: 60% said it was a "major" problem facing their age group.[27]

Concerns about the effect of technology on young brains—as yet unclear—have led both the World Health Organization and the American Academy of Pediatrics in recent years to issue guidelines that recommended no or limited screen time for children under five years old.

But the general obsession with our digital devices is only increasing. By 2017, according to emarket.com, US adults spent over 12 hours a day with major media. That included nearly 6 hours on digital devices (mobile phone, desktop, and laptop) as well as 4 hours on TV. A 2018 Nielsen study reported similar figures: It said American adults spend more than 11 hours per day watching, reading, listening to, or simply interacting with media.[28]

(Note that time in these studies is calculated based on time spent with each medium, including while multitasking. So, for example, an hour of watching TV while working on your desktop or laptop is counted as an hour of each.)

There are many ways to put some of that time to productive use—for starters, on training to get to a higher pay level or qualifying you for a better-paying occupation. Coding, for example, is a hot job category with too few applicants; a friend recently had to abandon plans for a fully financed startup just because she couldn't find enough competent coders. It's possible to teach yourself to code in 1,500 hours, and with the help of a mentor, you may be able to reduce that time to half or even just a third. Tech-related bootcamps can get you up to speed as a coder in under six months.

And, of course, there are career goals in fields other than technology that you can spend time pursuing. Many successful novelists have managed to write while supporting themselves at other jobs, among them William Faulkner (power plant worker), Anthony Trollope (33 years as a postal clerk), Fyodor Dostoyevsky (engineer), and Kurt Vonnegut (Saab dealer, among other positions). Joseph Heller worked in advertising by day and spent two or three nights a week working on a novel for eight years. "I gave up once and started watching television with my wife," he said. "Television drove me back to *Catch-22*." There's a lesson to be learned.[29]

When Malcolm Gladwell said in his 2008 book *Outliers* that "ten thousand hours is the magic number of greatness," many took that to mean that 10,000 hours of practice could make you a master of anything. Although the authors of the study he cited said he oversimplified their finding, nevertheless they pointed out that most people are not aiming to be world-class competitors but simply wanting to learn a skill (and certainly should be, if they're looking to get ahead). Further, they noted, everyone can improve performance by training properly for just a few hundred hours. Of course, the more additional training you put in, the better you will get.[30]

It is said that youth is wasted on the young. Generally, it's only as you grow older that you realize that the most precious commodity isn't money but time. You shouldn't spend it recklessly.

I often talk to my clients about the power of compounding. Investopedia explains it clearly. Assume you get a steady 5% return on your money. If you're a 25-year-old trying to accumulate $1 million by age 60, you'd have to invest $880.21 each month. If you don't start until you are 35, with the same goal and the same return, you would need to invest $1,679.23 each month. And if you don't begin until you are 45, with the same goal and same return, you would need to invest $3,741.27 each month.

When you fail to invest time in what you're doing, the power of compounding operates in the inverse: Your lost time is compounded. Once you fall behind your pack in moving on and moving up, you continue to fall further and further behind.

As Virtual Interactions Increase, Social Skills May Decline

The internet has hugely expanded the possibilities for staying connected—indirectly by email and directly by VoIP (Voice over Internet Protocol), which includes the ability to send instant messages and to text. Both are infinitely cheaper and quicker than the phone and mail systems that preceded them. It is easier than ever to stay in touch with your current network and to establish bonds with new people virtually anytime and anywhere.

But young people increasingly choose to spend time online rather than communicating with the people seated next to them. Instead of checking out one another at bars, they're checking their cell phone screens. In restaurants, they're sharing a table but not a conversation, instead sending messages with their thumbs. They have fewer experiences with personal contact in real time and in the same space, when subtleties like gestures, posture, and other body language as well as facial expression, pauses, and inflections are all part of the communication.

There is already some scientific evidence to indicate the pitfalls that result. One study reported that children who went briefly without technology were significantly better at reading human emotions than kids who had regular access to phones, televisions, and computers.[31] The idea that substituting Facebook and other second-hand emotional connections for direct personal contacts may affect your ability to process emotions is being investigated.

The *American Journal of Epidemiology* published a finding that using Facebook can even have a negative effect on your well-being. "While screen time in general can be problematic, the tricky thing about social media is that while we are using it, we get the impression that we are engaging in meaningful social interaction. Our results suggest that the nature and quality of this sort of connection is no substitute for the real-world interaction we need for a healthy life."[32]

This certainly helps explain why job candidates I interview often fail to make a good impression. Even when someone's résumé looks promising, when I try to engage him or her in conversation, the candidate often seems socially awkward. And in my business—probably in just about any business—that's a big deficit.

It's a deficit, in fact, in any situation where you're trying to make a positive impression, whether it's to initiate a romantic connection, to be welcomed to a community, to be considered for a membership, or to charm prospective in-laws. These circumstances and many others are all, at their essence, sales presentations, with the product being you.

A Harris Poll on job trends of 2019 found that "soft skills"—meaning interpersonal skills, communication abilities, and critical thinking—are important to 92% of employers in hiring decisions.[33]

Some people are just naturally good at presenting themselves in a favorable light, but practice benefits everyone, and practicing requires having actual conversations. When I ask candidates to tell me about themselves, I don't want them to repeat the information that they've already given me on paper. I want them to say something that can help start a dialogue or give me additional information and insight. Similarly, when I ask about their qualifications, including experience and training, I don't want a simple, "Yes, I have done that kind of work." I want them to add the extra information that makes them stand out from the rest, such as "and I overcame an interesting challenge," or "and I had some dramatic success," or "and I learned an important lesson," followed by details.

To turn a job into a career, moving upward is closely linked to high levels of empathy and likeability. Being competent or even outstanding at carrying out tasks isn't enough to be viewed as a leader. If your boss doesn't see you as someone who can communicate with others to inspire loyalty and motivate commitment, you won't become the head of the company, and you might not even make it to the head of your team.

Along with leadership skills, success also requires that you be a good listener. Part of that ability resides in reading nonverbal cues—not just facial expression but also body language, which is harder when a conversation is taking place onscreen.

And although working remotely saves you the hassle of commuting, you miss the opportunity of working in a group, where you can observe and take in all sorts of life lessons, from the art of persuasion to conflict resolution.

Also, the more time you spend in a virtual existence rather than an actual one, the more you're likely to miss valuable opportunities to create a network that is actually useful. How often has a LinkedIn connection, intending to be helpful but without real insight that comes from working side by side, "recommended" you for a skill you might not even profess to have? Facebook friends or multiple LinkedIn connections does not necessarily translate into a single sale or job lead. And it is not nearly as meaningful as having a few people who are sufficiently familiar with your skills and experience to turn you onto a good and appropriate job possibility and actually vouch for your work in a specific and meaningful way.

What social media does do is demand a lot of time. As with other investments, you should always be considering how much you are putting in vs. how much you're getting as a result.

Watching Devices and Multitasking Diminishes Performance

When Netflix CEO Reed Hastings was asked what he regarded as his company's chief competition, his response was surprising. He didn't mention Amazon, Showtime, or HBO, as you might have expected. Instead, he replied, "We're competing with sleep."[34]

An estimated 95% of us habitually turn to screens before bedtime, not only to access Netflix and other networks, but also to shop QVC, play games, read Kindle, or clean out the inbox (the tech equivalent of being the Sorcerer's Apprentice; it seems to refill with more mail even as you empty it).

Such stimulation causes increased electrical activity in your brain and makes your neurons race. And the resulting stress—not necessarily from something like an upsetting news story or unwelcome mail but just the tension of playing a game—causes the adrenal gland to release a surge of cortisone. This is not conducive to dozing off into restorative sleep.

In addition, if you habitually stay up late at night, your body readjusts your internal clock and you get "sleep-phase syndrome": eventually, you won't be able to fall asleep until whatever has become your new reset time. Midnight? Two a.m.? But if you've got a job to go to, chances are you can't reset the time you wake up. Result: you're chronically sleep-deprived.

Sleep deprivation has a lot of negative effects. It reduces your ability to learn (since it impairs attention, alertness, concentration, reasoning, and problem-solving) as well as your ability to remember (it affects your ability to "consolidate" memories acquired during the day). It increases the likelihood that you will make mistakes or have accidents. And it's linked to depression. Any of these is likely to impact the quality of your work.

So is multitasking. Many of us think nothing of keeping one eye on the cell phone and another on the task at hand. In the *Psychonomic Bulletin and Review,* the authors of a study on multitasking affirmed that it's well-established by cognitive science that attention has a limited capacity, and they confirmed what intuitively you might expect: There's a strong connection between attention and performance. Among the 200 people they sampled, 97.5% showed a significant performance decrease when they multitasked. Just 2.5% showed no decrease at all.[35]

Because increased exposure to technology has changed the way they think and act, it has been suggested that young people, often referred to as "digital natives," are adapting and becoming more adept at multitasking. However, a group of those digital natives—that is, some high school seniors—presented a research study at the American Academy of Pediatrics (AAP) National Conference & Exhibition that suggested that while Generation Z might be evolving to be very skillful at multitasking, most people performed best when focused on one task.[36]

(By the way, if you think you're a whiz at multitasking, take this online quiz: https://davecrenshaw.com/myth-of-multitasking-exercise/)

A few years ago, a study claimed that from the year 2000 (the beginning of the ubiquity of the internet) to 2015, the human attention span (the length of time to be able to concentrate on a task without being distracted) had diminished from 12 seconds to 8.2, This, amazingly, is even lower than the attention span of a goldfish, which someone (perhaps even more amazingly) had ascertained was 9 seconds. Critics of the study noted that "paying attention" really means being selective about responding to stimuli—thousands of which reach our eyes and brains (and also nose, ears, and skin) at any given moment. Being able to focus like that is critical. They say our problem is not a diminution of attention, but an increase in multitasking.[37]

Repeat studies duplicating the findings about the human attention span seem called for, but what is a fact is that the diagnosis of Attention Deficit Hyperactivity Disorder among children 4–7 is on the rise, increasing from 6.1% in 1997–1998 to 10.2% in 2015–2016 according to a report in the *Journal of the American Medical Association (JAMA),* in August 2018.[38] One observational study found that children who were lower screen users had lower rates of ADHD, but it wasn't clear whether kids with ADHD sought out more stimulation or whether the digital media exposure itself created increased risk for those symptoms.[39]

Dazzled by competitors in a memory championship, Franklin Foer, the author of *Moonwalking with Einstein,* decided to spend a year training himself to compete (which he did, and came out a winner). Among the many curious facts he revealed about how memory works, one in particular stood out to me: Before printing was invented, all people had to commit to memory every fact and story they had been told. Imagine the amount of information the learned class of the past managed to store in their brains to have available for easy access.

Today, we are not obliged to store a lot of information, since we can get images or facts from our phone or on the computer. Why learn the times table? Why do you need to remember the route to your friend's house? Or learn your cousin's phone number? Or commit a beloved poem to memory? Your memories are stored in your cell phone and warehoused in the Cloud. We don't keep in our heads the details of life that make us who we are. We are separated from our experiences, and we rely on our brains to do the mental exercise of trying to bring up a mental picture or do a calculation less often. We know what lack of exercise does to our muscles. Don't you wonder what it does to our brains?

Excessive Screen Time Affects Your Health

Medical costs typically suck up the largest share of your expenses as you age. The cost of care, insurance, and prescriptions is high, and your future needs are unpredictable. Being uninsurable, needing services or medications that insurance won't cover, or becoming too physically or mentally compromised to work can really blow a giant hole in your long-term planning.

A lot of medical problems are caused by sitting in front of a screen for decades, working during the day and watching entertainment at night. Some are minor, such as nearsightedness, dry-eye syndrome (because you blink less when staring at the screen), and poor posture. Some are not. What's more, when you're at your computer or watching TV, you are more likely to snack, more likely to be influenced to make poor nutritional choices, and less likely to move around—all of which can seriously and irreversibly compromise your health.[40]

And there's another, more subtle health risk from the light of your digital device, which has a higher concentration of blue light than natural light. Our body has a circadian system that synchronizes important functions, including digestion, body temperature, and cell repair, with a 24-hour cycle of light and dark. This activity is directed by specialized retinal cells that are very sensitive to light that is blue—the color of daytime sky. When they detect blue light, they mislead the circadian system, which has slowed down for night, into behaving as though day has arrived. Scientists also say we need darkness to produce melatonin. It's the hormone that signals the brain it's time to rest, so that the body makes physiological preparations such as muscle relaxation, a drop in body temperature, and other adjustments for sleep.

Disturbances in your sleep cycle and interference with the production of melatonin may cause diabetes, heart disease, stroke, and other problems with your endocrine, metabolic, and cardiovascular systems—and possibly (although the studies are inconclusive) even certain cancers, such as breast and prostate cancer.

Many of us have become more cautious about diet and exercise in the hopes of remaining physically healthy and minimizing the risk of expensive medical care. It seems that we should be cautious about our use of technology for the same reasons.[41]

Tech Distractions Reduce Productivity

When Pablo Sandoval, third basemen for the Red Sox, got busted checking out girls on Instagram in the middle of a losing game, he was benched. A lot of people have been distracting themselves on company time and getting away with it. In 2014, salary.com asked people if they wasted time at work and 89% said yes. While

62% wasted an hour or less daily, 16% admitted to two hours, 6% admitted to three hours, 2% wasted four hours, and another 2% wasted five hours or so each day.

But in the end, the consequences of wasting time on the job are as negative for you as for your employer.[42]

In my world, there's a lot of downtime: waiting for the phone to ring, waiting to hear about a fire we have to put out. I am sure it is also true of other jobs. In the past, when you had some downtime or just stalled out—lacking inspiration or baffled by a problem—you'd take a break just to be alone with your thoughts, and maybe one of those thoughts related to work would have some kind of payoff. Now you're more likely to distract yourself with a screen.

Of course, there have always been ways to distract yourself at work. When I started in the brokerage business, people would come in and spend an hour reading all the papers. Over time, they might extend that reading time to two hours, maybe three, instead of doing their work. I myself did the same thing. I'd read rather than cold call. One day, a manager walked over to my desk, pushed down the top edge of the paper I was hiding behind, looked me in the eye, and said, "Until there are a hundred people caring what your opinion is, you don't need to spend so much time reading." I had no rebuttal.

In fact, once I relegated reading to the weekend, I wound up being more productive than ever. I started calling execs early before their secretaries got in and monitored their calls, and I built my business. If I were a young broker today, I'm not sure I'd use my time as profitably. A Microsoft study found that 77% of people aged 18 to 24 responded "yes" when asked if the following statement applied to them: "When nothing is occupying my attention, the first thing I do is reach for my phone." Only 10% of those over the age of 65 answered "yes."[43]

More than 80% of workers have smartphones, and the same percentage of those people keep them in sight at all times. American consumers now check their smartphones an average of 52 times each day, according to the US edition of the 2018 Global Mobile Consumer Survey from Deloitte. That's up from 47 times per day according to the 2017 study by the consulting and professional services firm. About 35% of adults say they use their smartphones "very often" or "fairly often" for business calls outside normal working hours.[44]

Only 10% say smartphone use is decreasing their productivity, although 66% say they use the phone (at least) several times a day while working. But you have to wonder how many of them are doing something productive and how many are looking at pictures of burgers on Instagram?[45]

Employers surveyed in 2016 said they were losing two or three hours of employee productivity a day because of various distractions. Of the seven top reasons they cited, the first, second, fourth, and seventh were related to technology.

- Phone use and texting: 50%
- *Gossip: 42%*
- Surfing the internet: 39%
- Social media: 38%
- *Coworkers stopping by: 27%*
- *Snack or smoke breaks: 27%*
- Email: 26%

But decrease in productivity was only one of the problems related to distraction, according to the employers. The other consequences included compromised quality of work, lowered morale by others who have to pick up the slack, negative impact on the boss/employee relationship, missed deadlines, and loss of revenue.

Peter Cohan, a contributor to Forbes.com, estimates that people using Facebook while at work is cutting US productivity by 9.4%, resulting in $1.4 trillion in lost profits.

Some employers now use software that blocks access to sites that would be inappropriate choices during work hours, such as those focused on shopping, games, sports, or entertainment. But, of course, cell phone owners have the ability to get around that tactic—to their own detriment.

Of course, technology developers have some tricks up their own sleeves. In February 2018, it was announced that Amazon had won two patents for a wristband that tracks the wearer's hand moves and uses vibrations to prompt that you are performing a task incorrectly. It could even identify your pausing to scratch. However, it was unclear if Amazon actually planned to manufacture the tracking device and have employees wear it.

Distraction May Even Affect Your Personal Bottom Line

When you're distracted by technology, you're cheating your employer. But the consequences are also harmful to you personally. You're working longer hours to get less done. If you waste only five minutes an hour, that's an additional 40 minutes you have to add at the end of the work day. A lot of people are taking ten hours to do eight hours of work. It's like cheating on a diet. You're only kidding yourself.

Office workers aren't alone in losing income by not remaining focused. A waitress not paying attention to the patrons will probably make less on tips; a street vendor on his phone may lose a sale when a customer won't wait; a cabbie chatting on a call may fail to spot a fare, take a wrong turn, or have an accident.

If you weren't on your phone or browsing the internet, you could be making sure your report was great instead of good enough or kicking around ideas with colleagues that could benefit the company and improve your own prospects.

You can bet that people who become a huge success in life—the executives in the corner offices, the players at the top of the roster, the academics who get tenured spots—didn't waste time.

What's more, scientists now know that in between thinking about specific tasks, the brain goes into default mode called REST (Random Episodic Silent Thought) where most activity takes place in the most complex parts of the brain, and the mind wanders to review the past, consider the future, and make emotional judgments. This is when you can make life-changing decisions and come up with transformational insights.

What work are you not doing on a project that would help you shine and lead to a raise, what creative ideas are you not thinking about to launch a business or expand your opportunities and keep your freelance or consulting income coming? When you're watching a cat video, trying to find the perfect emoji, or checking baseball stats, what are you *not* doing that would have a positive financial impact on your life?

There's a saying in the hotel business that nothing has less value than the room you didn't book last night. When you lose a few hours binge-watching here or game-playing there, you haven't just lost a few hours. You have lost an investment opportunity in your own future that could have paid off big time.

2

Lifestyle Creep:
Keeping Up with So Many Joneses

More than a third of Americans admit their spending habits have been influenced by images and experiences shared by their friends on social media and confess they spend more than they can afford to avoid missing out on the fun. . . . They rank social media as the biggest "bad" influence when it comes to how they manage their money.

—*2019 Modern Wealth Index Survey*, Charles Schwab & Company

FOR MOST OF HUMAN HISTORY, the have-nots had very little idea of what they were missing. But 700 years ago, Marco Polo, an Italian merchant, unwittingly sparked a trend we know as "aspirational consumption." Even today top brands like Polo by Ralph Lauren owe their success to marketing to consumers' who long to own luxury products and hope that one day these high-quality posessions will be within their reach.

When Signor Polo led a galley from his home city of Venice into an unsuccessful battle against its rival Genoa in 1298, he landed in prison. To pass time behind bars, he recounted to his cellmate his 24-year journey along the Silk Road into China and Mongolia. The cellmate turned the stories into a book he called *Marvels of the World*.

The fabulous account introduced Europeans to many beautiful things, such as porcelain, raw silk, and ivory. These ignited immediate demand among 14th-century readers. Of course, since the printing press would not be invented until more than a century later, there were few copies of *Marvels of the World*. There were also few people who could read, and fewer yet of the kind of people who could afford the marvels described.

All that changed with the coming of Industrial Revolution, from 1760 to 1840, a transformative period of technological explosion second only to the present. Then as now, it created general prosperity and, as a result, the growth of a large population with lots of disposable income.

Helping things along, mass production techniques brought an abundance of goods. And advances in printing and the ability to reproduce images led to the emergence of magazines and ultimately to an entire new industry: advertising. That gave marketers a means to introduce tempting wares to a lot of people eager to make purchases that showed off their recently acquired wealth.

The first advertising agency opened in Philadelphia in 1841, and many more soon followed. Over time they developed increasingly sophisticated ways to whet buyers' appetites.

Ongoing developments in media technology over the following decades in radio, motion pictures, and TV have increased that appetite. Initially through radio programs, then in movie theaters and eventually in their living rooms, the general public heard about and saw how the upper class lived. They knew what kind of clothing and jewelry were in their closets, how they decorated their homes, which cars they drove, and what other indulgences they enjoyed.

Fortunately for both manufacturers and consumers, technology made it possible to produce more democratically priced versions of many aspirational desires. Sam Walton got rich giving ordinary folk the chance to buy the same kinds of things as rich people. (Others got rich in the same way; but Walton got richest.)

The Explosion of Trickle-Down Consumerism

These days, there are a huge number of wealthy folks to emulate, thanks to the introduction of "reality" shows like the *Housewives* franchises, *Lifestyles of the Rich and Famous* and documentary series that offer intimate peeks into the lives of people such as Whitney Houston, Britney Spears, Paris Hilton, the Osbournes, and the Kardashians. Initially, not all of them might have been truly famous or actual celebrities, but they were all definitely loaded, and unabashedly so.

Soon, through their example (or of others like them), wide swaths of the population learned not only to recognize Louboutin shoes, Rolex watches, and Louis Vuitton luggage but also to covet them and contemplate buying them, and sometimes actually doing so—even if they seemed way beyond the purchaser's pay grade. (I am thinking of a friend's 14-year-old daughter, who used baby-sitting earnings to treat herself to a $2,000 handbag.)

Researchers Marianne Bertrand (Chicago Booth School of Business) and Adair Morse (UC Berkeley's Haas School of Business) named the phenomenon "trickle-down consumerism." People tend to imitate others who are living the high life. What that costs them is not measured only in dollars. The researchers found that households that were more familiar with how rich celebrities lived reported more financial pressure. They also discovered a relationship between the income of the richest households and the number of personal bankruptcies in a state. In the current climate of increased consumerism, the researchers found, middle-class people had begun allocating some of their spending away from the basics—gas, utilities, and groceries—to items like clothing, jewelry, furniture, manicures, and exercise classes.[46]

Perhaps oddly, a recent psychological study found that it wasn't the snob appeal of having luxury goods that made people want to purchase them but rather that having such goods made them feel more accomplished.[47]

How Social Media Ramps Up Desire

In fact, our spending is far less likely to be directly influenced by celebrities than by the people with whom we're more likely to share tastes and, perhaps even more important, people who share our economic circumstances: our friends. The ubiquity of cell phones and the expanded use of social media in the last decade or two has exponentially increased the opportunities to get a really intimate peek into their private lives as well.

In 2018, reports Statista, more than 77% of Americans had a social media profile. Being on social media tends to expand users' networks tremendously. A circle of "friends" that a couple of decades ago might have been a one- or two-digit number might today easily run to three digits or more. The median number of Facebook friends in 2016 was 150. (Odd fact: 150 corresponds to something called "Dunbar's number," named for an anthropology professor whose studies of primates and human tribes suggest that human brains haven't the power to maintain relationships with a group much larger than that size.)

Friends' pins, pics, and status posts on Facebook, Instagram, Snapchat, and a host of other platforms put on display a huge, envy-inducing, acquisition-inspiring panorama of goods and services, many of which wouldn't even have been on viewers' radars just a few years ago. The display is purposeful.

It seems that many users of social media deliberately cultivate a glamorous image online. Two German studies of close to 300 college-aged Facebook users reported that people they identify as materialists use Facebook more frequently than other people, and that "as narcissists used [Facebook] for self-glorification, materialists use it to acquire and promote possessions."[48]

In fact, 20% of people admit to doing or buying things just based on how others will see it when it's shared, and 14% say they post something specifically because it looks expensive. All this has the desired effect. About 25% of people confess that seeing social media pictures of other people on holiday or enjoying new purchases makes them envious. Forty percent consider giving themselves a similar treat. More than 10% actually do.[49]

In one study, a whopping 90% of shoppers said that user-generated content—images, videos, text, and audios posted by people on social media—influences their decisions to make a purchase, outranking all other forms of marketing, including the runners-up: search engines (87%) and promotional emails (79%). And four out of five people are willing to pay more and/or wait longer for products that they've heard about through user-generated content.[50]

As you might expect, the youngest consumer base is especially impacted by social media. What one expert called the millennials' "financial Achilles heel" was fear of missing out (FOMO), which 55% admitted to, and 57% reported unplanned spending because of something they'd seen on their social media feeds.[51]

One recent study found that social media influenced 80% of Gen Z in their shopping and 74% of millennials, but just 58% of Gen X and 41% of baby boomers. Gen Z is very much influenced by Instagram (44%), Snapchat (21%) and YouTube (32%), while the numbers for millennials are significantly smaller (21%, 11% and 22% respectively).[52]

However, older generations are getting onto social media in greater numbers, and its impact on them has increased steadily. It seems inevitable that social media marketing will cast its shadow more evenly across all generations in the near future.

The collective influence of social media friends translates into staggeringly large sums.

In 2017 multiple news outlets reported that on average, 19% of a brand's sales—or *between $7 trillion and $10 trillion in annual consumer spending in the United States*—are driven by social conversations, both online and offline.[53]

Technological Innovation = More Stuff to Buy

Technology itself is becoming an ever-growing budget item. By 2013 it represented about 5.5% of a budget (even then, 1.5% more than savings), although at the time there were just a few items in the category: devices (computers and phones), high-speed internet connections, and music downloads.

Computer costs have actually gone down. In 2013, the sticker price for a new Apple Mac Pro with 128 GB of storage was $1,499. In January 2019 a comparable model cost $1,299. Phone prices have gone down as well. The global average price of a cell phone reached its high in 2011—$348.60—but was just $214.70 in 2019, according to Statista, the German online data-gathering source.

But although there are many more low-end models of computers than ever, buyers prefer models with more memory and faster speeds and features like interconnectivity with other devices, increased security, and better or bigger screens. And even though there are low-end phones as well, we're increasingly willing to pay more for what has become our everything-device for communication, work, photography and entertainment. With 75% of online video viewing taking place on phones, there is a demand for large displays along with improved camera technology, processing power, battery life and internet data speeds, though there has been some pushback against new $1,000 models.

Cable charges rose significantly from 2011 to 2015—a whopping 39%, almost eight times the rate of inflation.[54] And they continue to rise, generally 3% to 4% annually, says *Consumer Reports*, plus added charges for items such as "broadcast TV fees" and "regional sports fees."[55]

As a result, rather than pay for telecommunication, cable or broadcast bundled television services, many have cut the cord and gone instead with digital streaming platforms like Netflix, Hulu, and Amazon Prime and to TV networks like Fox and ABC that offer their own.

But the subscription model has its own financial pitfalls. You may initially sign up for only a few platforms. Then you sign up for Acorn just to watch one great new British detective series or Starz to catch a time-travel drama, and so on, and you keep those subscriptions going because you want more or because you simply forget to cancel after the free trial period. And you may create your own ever-larger, more expensive bundle over time.

Still, the main reason technology is eating up a bigger share of our budgets is simply because there are more things to buy—some obligatory, some optional. In the former category is the enforced subscription model that many tech providers have adopted. Microsoft urges customers to pay a subscription fee rather than buy the software outright. (Although over time the subscription will cost you more, the license can be used on multiple devices and comes with more features.) The printer manufacturers convince you to sign up for a monthly charge for printer cartridge supply and page-per-month usage. You are likely also to be paying for a bunch of apps you've signed up for over time. Some are useful, some not so much, and some you may even have forgotten about but continue to pay for.

You may also be paying for satellite radio and in-car video systems; for formerly free services that you upgraded to paid, no-commercial iterations; for extra storage on iCloud and Dropbox; and for domain names you may or may not actually be using.

Other technology-related expenses are about to become baked into your budget. From 2015 to 2021, sales of consumer wearables, like fitness trackers and smart watches, are expected to triple and sales of wireless speakers to increase by 25%. And despite declining cable TV subscriptions and the popularity of internet TV, large-screen, ultra-high-definition TVs are also anticipated to have tremendous growth because we consumers have become used to multi-screen viewing. (Our smartphones trained us.)

And we'll increasingly succumb to the lure of Internet of Things (IoT) technologies—the interconnected devices and home appliances meant to save money, time, and energy. By 2015, just 10% of smart homes had a security system, 9% had a smart thermostat, and a smaller number had lighting automation, home theater and entertainment, electronic locks, audio/video cameras, and automated shut-off valves.[56] By 2018, 90% of US consumers owned some form of smart home device and nearly 70% had a voice-controlled systems (like Amazon Alexa or Google Home.)[57]

A lot more of us are projected to get these as well as other IoT items that will come to market, such as self-driving cars, refrigerators that automatically order groceries, and possibly even a home automation machine that vaporizes your favorite scent.[58]

Planned Obsolescence and the Upgrade Mentality

We're not only buying lots of new things; we're very quick to replace the old ones.

My father likes to recall how when our family home was being constructed in 1962, he brought his father over to take a look. Accustomed to homes built of plaster over a frame, my grandfather was incredulous when he saw men working with Sheetrock®, a layer of gypsum compressed between two layers of paper. "You're spending $29,000 on a house with paper walls?" he asked.

"This house will be around long after you and I are gone," my father assured him. True enough; when my grandfather passed away, the house was still standing. My father is still living in the house, whose lifespan will likely exceed his as well.

First manufactured in the United States in 1917, Sheetrock® was widely adopted within a decade when builders found that Sheetrock® walls could be installed more quickly than walls of plaster over frames (which had to dry between coats) and required less labor. By reducing construction costs by about a third, it made housing way more affordable.

Few would have thought that the technology that produced Sheetrock® would be so impactful, enabling the building boom that led to urban sprawl. It was only one of an unprecedented number of technological advances in the last century or so that transformed and also greatly improved virtually every aspect of our everyday existence. Much of the US population has in its homes all 60 of what Wikipedia today describes as "basic" home appliances, from air conditioners to fan heaters, garbage disposals to microwaves, mousetraps to televisions—all of them innovative technologies at the time they were developed.

There's a bit of a catch, though, for the providers of innovative goods. Once a new technology has been introduced and widely adopted, sales growth often slows down. Take the automobile, for example: Its original purpose was to get you from Point A to Point B faster than by other means. Those benefits were so clear that it quickly found a market.

After the original models were introduced, subsequent versions became even faster and more comfortable. What's more, they were built to last. When Henry Ford had perfected the design of his Model T, he saw no reason to change it. Of course, that meant that when the market was saturated, there would be fewer prospective customers. This put manufacturers in a quandary.

A few years later, in 1924, Alfred Sloan, the head of General Motors, came up with (or, as some suggest, usurped from the bicycle industry) a wily idea: "instilling in the buyer the desire to own something a little newer, a little better, a little sooner than is necessary," as industrial designer Brooks Stevens put it.[59] Whatever the idea's origin, Sloan implemented it to stupendous effect.

That was a particularly daring move during the Great Depression, when most people were struggling just to make ends meet. Boldly, Sloan had his company bring out new models every season, cars that could be perceived as "upgrades" that would appeal to buyers who wanted to display their own growing economic strength.

He also decided not to have his brands compete with one another. Instead, he established a hierarchy of pricing structure, from the least costly Chevrolet to Pontiac, then Oldsmobile, next Buick, and, at the acme, Cadillac. As he had hoped, customers stayed loyal to GM, and, as he had envisioned, they progressed up a purchasing "ladder" as their buying power grew.

That planned obsolescence strategy is now on steroids. By tomorrow, today's "new" is old and its "better" is obsolete. Manufacturers have become increasingly artful at persuading us to frequently replace the products they have sold us with successive versions for reasons that amount only to what economists describe as "marginal utility."

Utility is the satisfaction or benefit you get from consuming a good or service. Marginal utility is the change related to an increase or decrease in the consumption of that good or service. The "law of diminishing marginal utility" suggests that the first "unit" of consumption gives more utility than subsequent units, and each successive unit gives you less.

My grandparents loved to brag about having the same toaster for forty years, but today consumers who already have a toaster are likely to buy a replacement simply because it's—well, different. At one time, you had three or four toaster choices—black or chrome? slots for two slices or four?—but now you've got an aisle full of niche-specific models, each vying to make you feel it's a must-have.

Perhaps there's a point to the existence of a toaster that simultaneously toasts bread and boils eggs; one that toasts muffins and poaches eggs; or one that also toasts rolls and grills hot dogs. But is there really a need for a transparent, portable, or hand-held toaster? A $300 model decorated with Swarovski crystals? Or that can burn decorative images or even a short note in your handwriting into the bread slice? How does any of these add utility to a device whose purpose is really just to make toast, an item meant to be consumed and forgotten within minutes?

They don't. But we're suckers for anything "new" and "improved."

Our Culture of Disposability

By tempting customers with newer or better versions of what they already have, planned obsolescence is still an effective way to generate sales. However, planned obsolescence is not only a matter of bringing in the new but also of phasing out the old.

Repair shops, once a staple along any Main Street, have virtually disappeared, since you're often obliged to buy a new piece of equipment if your current one fails. Even if you prefer the older, simpler (and often sturdier) model, replacement parts may not be available. And once you're forced to replace the old model, you may discover its accessories, formats, and features are incompatible with the newer one.

How many versions of your favorite music have you bought as successive generations of recordings replaced one another? Your parents' 78s were superseded by 45 RPM singles that in turn were superseded by 33 RPM vinyls and then by 8-tracks and cassettes. Then came CDs, which became obsolete in a couple of decades and were replaced by music downloads. Now streaming has becoming the music delivery source of choice.

And how much have you paid to transfer 8mm films to videotape and then to DVDs and eventually to disc drives?

While telephones of my childhood lasted for decades, the chargers that fit our two- or three-year-old cell phone didn't fit the newer model, so we needed either new chargers or a new phone. Phone customers now are offered an automatic upgrade plan that guarantees the latest model of their phones and commits them to pay for it before they even know if they'll want it. But they may have to buy it anyway. The old operating system likely won't work with the newest software.

Manufacturers of products that fail quickly because of slipshod design, poor construction, or inferior materials are also, some darkly suggest, deliberately creating "planned obsolescence."

They point, for example, to the light bulb designed by mastermind electrician Adolphe Alexandre Chaillet. Having dangled from a lighthouse ceiling in California since 1901, it made it into *The Guinness Book of Records* in 2015 after emitting light for 114 years—over a million hours. Some say Chaillet's carbon filament was just too expensive, but skeptics ask why the lighting industry was so slow to come up with long-lasting, energy-efficient, cheaper-in-the-long-run alternatives to the short-lived incandescent bulbs we used for over a century? In fact, a scam was uncovered that showed several international companies had colluded to artificially reduce light bulb lifetimes to 1,000 hours.[60]

Companies can also force consumers to replace equipment before it's worn out by making maintenance impossible, a situation that sparked the "right to repair" movement. It started with farmers, who rebelled against the long waits for manufacturer-approved sources to repair their equipment. As a double whammy, not only were they unable to obtain parts so they or others could do the work, but also such action would void the warranty.

As computerization became a vital part of car manufacturing and manufacturers controlled the information necessary to repair or supply parts, automobile buyers also began to protest. In response, in 2013 Massachusetts passed an automotive Right to Repair bill. In 2014, vehicle manufacturers' associations agreed to comply with its provisions in all 50 states.

A parallel issue exists with software and electronics buyers who sign purchase agreements that mean they don't really "own" the stuff in the sense they might believe, and the provider limits their rights to modify or repair it. In response, The Repair Association (formerly the Digital Right to Repair Coalition) was founded in 2013 as a member-driven nonprofit with a mission to support the high-tech repair, reuse, resale, and recycling industries.

Following 2017's "Batterygate"—when Apple was caught throttling phone processors in devices with run-down batteries (but not informing customers or allowing others to fix those batteries)—the company responded with several solutions. More important, the incident inspired widespread "right to repair" activism promoting legislation that would allow consumers access to parts for and the right to repair items that have embedded software.

But just because we have the option to fix things doesn't mean we will. We should consider our complicity with the manufacturers who encourage us to replace and upgrade.

"We Have Seen the Enemy, and He Is Us"

In May 2017, The German Environment Agency, noting that most electrical appliances were being discarded after an ever-shorter period of time, published "Strategies Against Obsolescence" that defined four types of obsolescence: material (defects), functional (incompatibility between replacement and older parts), economic (replacement costs little more than repair), and psychological (desire for a new item despite the fact that the old one is still functional). Strategies for the last included making designs timeless and modular—easily updated at the technical, functional, and aesthetic levels so they can remain attractive both for their utility "and as a social symbol."

By successfully persuading customers to replace older items with a newer model, many companies have produced handsome returns for stockholders. But there's a saying that if you're in a poker game for half an hour and haven't figured out who the patsy is, it's you. As an investor holding Apple stock, you might hope that all the owners of iPhone 7s will replace them with iPhone 8s as soon as they're released, and then replace 8s with Xs, and so on. But as the owner of an iPhone 7 that suits your needs, wouldn't you be smarter to stay with it?

Philosophers have recognized that a claim that an idea or proposal is correct or superior *exclusively* because it is new and modern—that is, based on its appeal to novelty (also called *argumentum ad novitatem*)—is a fallacy.

But the prescient Alfred Sloan of GM realized what anecdotal and research evidence seems to confirm. We are simply drawn to novelty: newer and more expensive iterations, with optional, possibly irrelevant, features. In short, we're just buying this stuff because we want it.

The Irrationality of Manipulated Desires

We have an amazing capacity to attach plausibility to expenditures that defy logic. That "we" includes me. My wife and I paid $250 extra for our dishwasher to get an element that shot water with a pressure of 7,000 pounds to clean pots and bakeware—even though in our household no lasagna pan or casserole dish goes into the dishwasher before it's thoroughly scrubbed.

And, of course, I see many variations of this tendency when I work with new clients to review past spending and consider adjusting their budget for the future. People try to justify purchases that without doubt fit the definition of "marginal utility." Instead of buying something just as functional that costs a lot less, they purchased the more expensive version. "I know that BMW might not have been such a practical choice, but I just loved that car," they'll say. "I felt so great driving it."

"Now that it's three years old," I might reply, "do you still feel great about it?"

It doesn't take long for one year's "must-have" to become next year's "meh."

One couple came in shortly after returning from a $25,000 vacation that they described as "the trip of a lifetime"—mostly, I suspect, because it cost more than three times what they usually spent. They had probably convinced themselves to go for it with some version of "You only live once" or maybe "Life is short" or "My father died young." If I had $100 for every time I've heard any of those remarks, I could fund my own trip of a lifetime.

The irony is that most people come to me because they're concerned about being able to maintain their lifestyle for a lifetime as they erode that possibility with splurges. "I don't normally spend this" is the gateway phrase to lifestyle creep.

Many of us splurge when we can't afford to. And some of us, when we do have money, even a lot of money, manage to keep spending until it's gone. Maybe that tendency to indulge is built into our DNA. Our ancestors foraged for food and primitive hunting tools, and food was rarely abundant. When it was, they gorged themselves. The stores they built up helped them survive periods of scarcity. In a period of abundance, that primitive response isn't so useful. It leads to obesity—and perhaps it explains why we overspend when we're feeling flush.

In 1955 the British economist Cyril Parkinson formulated Parkinson's Law: Work expands so as to fill the time available for its completion. I would like to add Kamen's Conclusion: "Lifestyle creep expands so as to surpass the amount of money available to satisfy it." There are countless tales of superrich celebrities who wound up in bankruptcy court.

In the 90s, the baseball player Jack Clark filed for bankruptcy in the second year of a three-year, $8.7 million contract with the Boston Red Sox. He listed $6.7 million in debts. He owned 18 automobiles, including a 1990 Ferrari and three 1992 Mercedes Benzes. His lawyer, Fletcher A. Robbe said, "He had some expensive hobbies, and I think they got ahead of him."[61]

The paychecks have gotten way bigger over the years for sports and entertainment figures, which for some just means more to lose. Mike Tyson, who earned about $400 million in his career, was $27 million in debt by 2003. (Taxes, divorce payments, $300,000 in limousine costs and Siberian tigers do add up.) Between 1996 and 2011 alone, Nicholas Cage reportedly banked $150 million, but he spent all that and more on shopping lists that included a couple of castles, his own Caribbean island, a private jet, and a collection of shrunken pygmy heads. And it boggles the mind to figure out how Evander Holyfield could have blown a reported $230 million.

A recent study reported that nearly one out of six former sports players—whether stars or not, and irrespective of career earnings or length of career—go bankrupt within a dozen years of retirement. It's become such an issue that team owners, who are acutely aware of the problem, are now making financial planning a routine part of offseason training programs.

Lifestyle creep comes in many forms and will create a long-term problem for all generations—from millennials to baby boomers—although their reasons for not addressing the problem may be different. Young people tend in general to live in the moment, even though their generation's prospect for retirement funding through Social Security aren't certain. And those closer to retirement may become less cautious about spending once their basic expenses decrease when the kids are out of school and the house is paid off.

You can see it in big ways (country club memberships, more eating out, more luxury trips) and in small ones (coffee-making machines, wearable fitness trackers, storage units to store all the stuff they've replaced with newer but questionably more beneficial models).

You might be unable to afford a Siberian tiger or even a trip to the Safari Off-Road Adventure at Six Flags but succumb to multiple minor indulgences because you feel entitled. If you're going out for hamburgers and see the picture a friend posted of himself enjoying a fancy steakhouse platter, you may think, "Why am I treating myself like a second-class citizen when Jake's eating steak? I work as hard as he does. I should have steak, too." Why not?

Here's why: You're the ultimate gatekeeper of your own financial security. You'd be better off maxing out your 401K instead of trying to one-up your social media network. If your plan is to enjoy now and pay later, little by little, you're squandering your future. And the fact that cashless transaction are on the rise just makes it easier.

3

Going Cashless

The Abstraction of Spending

It's amazing how fast later comes when you buy now.

Milton Berle, comedian

Scene: Checkout line at Target. Mother pushing stroller containing child who appears to be less than four years old. He is holding a toy he apparently just plucked from a shelf.

Child: I want it! Buy it for me!

Mother: Sorry, that's just not going to happen.

Child: But I want it.

Mother: Sorry.

Child: Then I'll buy it for myself!

Mother: You don't have any money.

Child: I have a card! In my wallet!

A HUSBAND AND WIFE, NEW CLIENTS, were meeting with me to discuss a typical concern. They weren't setting aside enough for retirement. We began to review their situation. They told me their joint income approached $500,000, and when I asked the approximate value of their assets—their home equity, savings, 401K investments and so on—they quickly came up with a number. I wasn't surprised. Generally, prospective clients are able to estimate their assets within a couple of percentage points of what we find when we go through the step-by-step review.

But then I asked them to estimate their annual spending—what it costs to be them, on an annual basis. They quickly calculated their big payments (such as mortgage and car) and their usual credit card balances, but they really weren't sure about the rest. We scheduled a second meeting, and I told them to take some time to review two or three months' worth of bank account and credit card statements and calculate their spending by categories.

When the couple returned, they told me they were shocked by what they'd found—how much they'd been spending on restaurants, how much they were paying for memberships and subscriptions they didn't need, and so on. Like many people, they had opened a bunch of taps that were draining away money without their really noticing it.

A great many new clients have no idea of what they've been spending. Often, the ballpark figure they give me at an initial meeting isn't even 10% of what they spend in a year. Once they look at the numbers, before I even utter a word, they themselves say some version of "No way. We 're not going to be able to keep this up and retire happy." Sometimes the most important part of our planning is a direct result of the insight they get from this process.

Over time, it's easy to accustom ourselves to routine expenditures that in the past might have cost many fewer dollars or taken a much smaller share of the budget. I have mentioned one such item—technology—that barely registered a couple of decades ago. The category is unusual, since many of the line items exist only virtually: cloud storage, internet access, website maintenance, cell phone service, streaming music, downloaded videos and books, and the interconnectivity costs of the Internet of Things.

The options are expanding, and our reliance on technology will only grow. But the fact that we're allocating so much money to things that don't exist in any three-dimensional sense is a reflection of how our world is increasingly virtual—as is our relationship with money itself.

Removing the Pain of Paying

Not long ago, at the end of every one-week or two-week pay period, American workers, from blue collar to professionals, got a physical paycheck. They'd stand on long bank lines on payday, deposit the check, and withdraw some of it as cash.

Now many of us make a deposit by waving our paycheck in front of a smartphone. Or maybe we don't even see the paycheck, since it's deposited directly into our accounts. In 2019 Visa and PayActive announced an agreement to introduce an even newer concept—offering real-time access to wages the very day you earn them. PayActive partner employers can make deposits into employee's accounts, and employees who have downloaded the PayActive app can choose to access the funds, which are linked to a debit card or a new PayActive Visa prepaid card that's accepted by merchants or at cash machines and points of sale.

Paper currency plays no role in these transactions. You don't turn a paycheck into cash, and you don't often pay with cash, either. I've often put a couple of hundred dollars in my wallet and realized, weeks later, that most of it was still there. I surely spent many times as much as the cash in the interim, but I couldn't make a quick calculation of how much because I didn't notice it going.

When you don't actually see your money disappear, it's easier to part with it. Businessman Frank McNamara made a fortune by capitalizing on this fact.

One day in 1949, when dining in a restaurant, McNamara forgot his wallet and found a career. Although his wife paid the tab, Frank determined that he would never again be embarrassed by not having cash to settle a bill. A year later, he returned to the restaurant and paid for dinner by presenting a product he had devised: a piece of cardboard he called a Diner's Club card. Within a year, his fledgling company had 10,000 cardholders, and 30 businesses accepted his cards.

By then, some individual stores had issued what they called "charge cards" to regular customers (whom they would bill monthly), but the Diners Club card was the first that could be used across a variety of businesses. McNamara was able to monetize his idea by convincing the business owners to pay him a fee of 7% against the charges racked up, based on his prediction their business would increase because cardholders would spend more than non-cardholders.

When his hunch proved correct, other companies jumped on the card bandwagon. Bank of America was the first to offer revolving balances, meaning the credit limit is set by the lender, but you chose how much you will charge up to the limit and how much you'll pay each month. It went so far as to mail 60,000 activated cards to people in California. Though that promotional stunt cost the company millions in fraud and delinquencies, Bank of America persevered. It sucked up the costs, issued even more cards, and soon turned a profit.

McNamara's genius was in identifying and capitalizing on what's now established as "the credit card premium"—the fact that we'll spend more if what we're spending doesn't *seem* like money. Casino operators, who force you to exchange cash for chips, have been taking advantage of this tendency for a very long time. Though poker chips came into use in the 19th century for several reasons—among them was the colors that made them easier to tally than coins or currency, and using chips prevented players from grabbing back cash bets—the most compelling was that chips encouraged people to gamble more freely.

If you were peeling off $100 bills to place your bets rather than tossing in chips, you'd be far more aware of what you were losing and probably leave the table sooner. For that same reason, almost no slot machines accept coins these days. You put your twenties into a machine, and when you walk away at the end, even if you're a winner you get only a voucher, not cash. At every step, they try to separate you from dealing with actual money.

The less real your money feels, the easier it is to part with. Dan Ariely, a behavioral scientist, explains that chips, loyalty cards, and credit cards all remove what he calls "the pain of paying." In the old days in Britain, when you needed gas or electricity, you literally had to put coins in a meter in your home to pay for it. As a result, I'm pretty sure people were more careful about turning off a light and turning down the heat when it wasn't necessary. (And deferred pain is always preferable to current pain. That's why even though you know that eating sugar may ultimately lead to obesity and other health issues, you may still order dessert, and if you're paying the electric bill at the end of the month, you're more likely to leave the lights on.)

Online gambling, already a $45.8 billion-dollar-a-year business in 2017, was anticipated to grow to $94.4 billion by 2024.[62] Part of the reason, as *Tech Addition Magazine* warned, is that it lacks some of the constraints that can keep casino betting in check. If a person goes on a two- or three-day binge in a casino, the person's absence might be noted; but no one knows if you're gambling at home. It's also much less of a challenge to stay away from a casino or other betting parlor than it is to avoid your home computer. (You're also less likely to spot fraudulent practices when you're playing online.)

But what really makes it dangerous is that you don't need to physically handle anything that looks like cash, not even a credit card. Access is just a matter of hitting a key, which makes impulse betting and chasing losses easier. On the internet, it's easy to forget that *electronic money is real money.* You don't know when you've crossed over from virtual spending to virtual overspending.

Cash Is Losing Its Currency

Thanks to ever more sophisticated technology, the future of currency is more and more uncertain. Starting in the 1990s, electronic banking accustomed people to non-cash transactions and settlements, and they adapted to the changes rapidly.

In a few countries—including Malaysia, Saudi Arabia, Peru, and Egypt—99% of transactions are still in cash, but globally the trend is otherwise.

The Scandinavian countries have been leaders in phasing out currency. Since 2015 the Central Bank of Sweden has been pursuing the concept of an e-krona, a digital currency. In late 2018, it released a report that it is beginning to design a technical test to see which solutions are practical and work on ways to clarify the legal standing of e-money, among other issues.[63]

Cash today represents just one percent of the Swedish economy. Even children in Sweden often use credit cards. More than half of the branches of the country's biggest banks don't accept cash deposits or keep cash on hand. Ikea opened a cashless store in Sweden because so few people were paying with cash, and handling the cash transactions was not cost-effective. The most problems cropped up in the food court, and sometimes the customers who had only cash were simply waved through.

Having discovered that 60% of the value of US currency was in hundred-dollar bills, Kenneth Rogoff, former chief economist at the International Monetary Fund, makes a different kind of case for eliminating paper currency in *The Curse of Cash.* He argues that it helps support corruption overseas, promotes tax evasion, and encourages illegal immigration since it's used to pay undocumented workers. In addition, he says, eliminating currency would also reduce theft, prevent dipping into the till, and even cut the spread of disease. (Currency is very unsanitary.) He recommends getting rid of larger bills first and then letting the smaller ones fall into disuse.

Some countries have already begun the process. In 2000, Canada got rid of its $1,000 bills, and Singapore ditched its $10,000 bills. In the UK, there have been suggestions to scrap the £50 note.[64]

In November 2016, Prime Minister Narendra Modi of India invalidated 86% of the country's currency, in part with the hope that those who stashed away illegal cash wouldn't be able to exchange it for new notes at banks. But that didn't pan out. There was a 99.7% return of the banned 500 rupee and 1,000 rupee notes—the highest value currencies before demonetization. The secondary purpose, which was to encourage digital transactions in a country where 90% of transactions had been in cash, has been far more successful. New electronic solutions gave a boost to digital alternatives, and airlines and many merchants have embraced QR codes, the square boxes that can be scanned into mobile devices and work like bar codes but contain more information.

China, too, has been very successful in making the transformation to digital payments. In 2013, everyone in China was paying cash, but within just four years, they had overwhelmingly switched to paying by phone. The way it works is that you link your bank account to a mobile app like SEQR (pronounced "secure" and used widely in Europe). To make a payment, you scan the vendor's QR with your phone, and enter your pin number. The money is transferred, and you receive a receipt online. In China, churches accept donations by app, and street musicians and beggars post boards with their QR code. By 2017, mobile payments in China amounted to $5.5 trillion, roughly 50 times the size of America's $112 billion market.

In fact, the huge number of Chinese tourists visiting Japan (eight million in 2018, who spent an estimated $14 billion) is creating pressure on that country to become more cashless. A Japanese government estimate in 2016, the most recent data available, found that credit and debit cards were used in only about one in five consumer payments there.

A number of businesses in New York have stopped accepting cash altogether. They say transactions can be handled more quickly, and they don't have to spend hours counting currency. More significantly, cash had become an insignificant part of their business—as little as one or two percent, estimated one restaurateur.[65]

There's some pushback of course. The city of Philadelphia passed a bill that as of summer 2019 would require retailers to accept legal tender. Sponsors of the bill had argued that going cashless was unfair to people without bank accounts or credit cards and to the traditionalists who just wanted to pay for their coffee with dollar bills and coins. Rental businesses that require deposits, parking lots, and other businesses were exempted. In April 2019, the salad chain Sweetgreen reversed its card-only policy, stating that it unintentionally excluded those who preferred to—or could only—pay with cash.

But these efforts may ultimately prove fruitless. By 2016 more than six in ten Americans agreed that "the United States will be a cashless society, in which all purchases are made with credit cards, debit cards, and other forms of electronic payment."[66]

And a Pew Research Center study released in December 2018 reported that about 29% of adult Americans in a typical week make no purchases with cash, a 5% increase from 2014. The percentage who say they use all cash went down to 18% from 24% in 2015. People with incomes over $75,000 are the most likely to use little cash, but older people reach into their pockets and wallets to pay.[67]

I recently took the three-minute shuttle train in New York that goes between Grand Central Station and Times Square. A street performer followed a brief guitar performance with a plea for tips, acceptable in cash or via either of his mobile apps, Venmo and Square Cash. The revolution has definitely begun.

No wonder Tim Cook of Apple has predicted that the next generation of children "will not know what money is." That may happen more quickly than we might have guessed.

Separating the Act of Spending from Its Consequences

In a prescient 2006 IBM commercial, a man roams through a store, plucking multiple items off shelves and out of refrigerator cases and stuffing them in his coat pockets. Just before he passes through the exit door a guard stops him, but only to remind him that he hasn't taken his receipt.[68] It turned out that thanks to a Radio Frequency Identification chip (known as an RFID) that had been implanted at the base of his thumb, the items he appeared to be shoplifting had all been scanned and charged to his unique RFID once he put his hand on them.

By late July 2019, at least 4,000 Swedes had paid $180 to have rice-sized microchips implanted at the base of their thumbs in order to go walletless and pay for food, purchase a train ticket, and enter keyless offices with a touch or a wave. If the technology proves workable and efficient, it seems highly likely that retailers will over time phase out other systems of payment.

One big downside of cashless transactions is that they require surrendering a great deal of privacy. The nation state and third-party providers have complete control of transactions, individual use of money, information about public monetary assets, and interest rates. And electronic systems can be hacked and then fail. In fact, even the Swedish government advises keeping a cash stash in the event of war or some other widespread calamity. There is also a possible physical consequence of having radio frequencies aimed at our bodies every time we buy a pack of gum.

But equally important, from the point of view of this book, is how such a system—which dispenses completely with any tangible reminder of payment—will go even further to separate the act of spending from the consequences.

Amazon might actually transform shopping with a sophisticated and less cyborg-reliant solution that it unveiled in early 2018 when it opened Amazon Go. You just pluck items off the shelf and put them in your bag (a reporter noted that it feels like shoplifting), and thanks to some wizardry performed by a bank of tiny overhead cameras that see and identify every item in the store, what you buy is automatically charged to your Amazon account. You leave the store with your bags, and Amazon sends you the receipt.

Amazon was planning to open 3,000 cashless stores around the country by 2021, but in April 2019, because of pushback, the company agreed to accept cash at its physical stores. Still, it seems the trend to cashless may ultimately prevail.

For consumers, the advice is caveat emptor. "For now, visitors to Amazon Go may want to watch their purchases," noted reporter Nick Wingfield after visiting the initial cashless store. "Without a register staring them in the face at checkout, it's easy to overspend." Yup. That's the idea.[69]

Credit Is a Gateway to Burdensome Debt

There used to be more barriers to accessing money when you needed it. You couldn't get a loan unless you had a job and were bankable. You couldn't get a credit card unless you were worthy of credit, an assessment determined by a bureau that kept tabs on your payment history.

But online lending—whether for an actual loan or as part of a credit transaction—has made it easier for people to take on debt that they can't manage, a temptation that people often succumb to when depressed. (Or mentally ill. "In the past, if you'd walked into a bank and asked for a $50,000 loan and the banker asked why, you could say, 'Because the world is going to end,' and then you could prove you had no mental capacity, and that debt could be written off," said Polly Mackenzie, director of The Money and Mental Health Institute. "But if you do it online, they can't know, and there is no reasonable proof.")[70]

Credit card companies are relentless about courting new customers. Among the most problematic of their strategies is sending a flood of credit card offers to new high school graduates. The companies make a bet on the premise that anyone who graduates from high school and gets into college will make money. Seeing these people as potential customers, they give them credit. (Perhaps they should do some interviews with parents of recent college graduates first.)

Initially the credit line may be low. But once a card user makes a habit of regular payments, the credit limit is raised automatically.

In fact, if you've tried arranging for a card with a limit on it—for a teenaged child, for example—you've no doubt found it practically impossible. The companies love increasing the limit. The more you borrow, the better for them.

That's why your chain isn't yanked when you go negative. Deficit spending is as easy as non-deficit. Yeah, if you overdraft, they ding you with $34 *per transaction*, but you don't viscerally feel the difference between an account that has $400 in it or one that's minus that amount; you'll just owe additional interest on the amount that got you into the hole.

Wildly popular credit cards give you lots of travel points that seem enticing. But the cards come with high annual fees, and unless you're very disciplined about paying off your balance, you may have buyer's remorse pretty quickly. My client's daughter used her credit card to pay her rent and get lots of points but somehow neglected to deposit that rent money back into her account. Within months, she had a five-figure debt plus high interest. She ruefully admitted that whatever tickets she got with her reward points would have cost a whole lot less than the interest she paid working off her debt, which included the hefty card membership fee.

The more irresponsible you are with your finances, the more profitable it is for the financial industries. If you have left a balance of $48 over the amount you've paid in one cycle, by the next it may have grown to several times that much, including penalties and additional charges. And even as the number gets bigger and bigger, you can just keep banging the card. You feel bad about the thousands you owe, so you'll at least make the minimum payment. That allows you to drown one month at a time.

People tend to use a card more often and more willingly than cash, and when they use it to shop, they will also consider higher-priced items. Researchers have found that people paying with credit cards tend to think more about the features of a product, whereas those paying with cash focus more on costs (for the product itself, shipping, assembling, and so on).[71]

If there's a minimum purchase required to use a credit card, a business may benefit. Since many of us don't carry around much cash these days, we may wind up spending more than we had intended in order to meet the qualifying amount, the same way we purchase more when a "sale" item is marked "must buy two." That's a win for the retailer—and likely for the credit card company as well. If you don't pay off the minimum balance, you increase the size of the loan. Incremental amounts may be minuscule, but they add up.

By 2019 almost 60% of Americans held at least one credit card and on average held 2.5. The higher one's income, the higher one's average credit card debt (in part, of course, because creditors are even more eager to lend you money if you're a high earner). Those in the age group 45–54 had the highest credit card debt ($9,096), and the age groups bracketing them—people 35–44 and 55–66—had near identical average debts of about $8,200.[72]

There are many estimates that vary from these figures, but what really matters, of course, is the number of credit cards you personally are holding and the rate you're paying if you don't pay in full at the end of each billing period.

The rates are outrageously high. They average 19.24% for new offers and 14.14% for existing accounts, but they vary enormously though they're generally somewhat lower for prime borrowers, people with good credit ratings. The rates are so high because in 1978 the Supreme Court let credit card issuers "export" nationally whatever interest rate was allowed in the state where they were headquartered. Some states (among the first, South Dakota and Delaware) wanted to attract financial companies so they dropped their usury laws, which is why you might be mailing your credit card payments to those states.

Fortunately, card companies are now at least required to be transparent about the rate you're paying to carry a balance.

Figures for total credit card debt in the United States seem to be all over the map, but one estimate is that revolving debt stood at $1.04 trillion at the beginning of 2019. That's up from $857 billion in 2013.[73]

By issuing cards, banks are in effect making loans for your purchases, loans that might represent only pennies in profit. But multiply a penny by 10 billion—the number of credit cards currently estimated to be in existence worldwide—and you know why the companies are doing so well. (According to its 2018 report, Visa took in $20.6 billion in 2018 with net income of $10.2 billion. For Mastercard in 2018 the respective figures were about $15 billion and $5.9 billion, and for Discover in 2017 it was $9.6 billion and $2.1 billion.)

Analysis by Elite Personal Finance estimates that FDIC-Insured Commercial Banks and Savings Institutions earned roughly $108 billion in credit card interest in 2018—an increase of $16 billion over 2017.

Paperless Billing: Saving Trees But Costing You

The virtualizing of bill paying, a combination of automatic payments and paperless billing, may also result in your spending more. It's much more likely you'll skip past an email with your monthly statement attached than you'll overlook a bill that comes in your mail.

If you don't have automatic payments and fail to send in your money, you can be subject to late fees or worse—cancellation of your car, health, or life insurance. (To reinstate a life insurance policy, you will need to submit to another medical exam, and very likely you will be unable to get the same low price a younger, healthier you had contracted for.)

If you do have automatic payments, you can get lazy about checking the statement. You may find that your credit card company is collecting only the minimum it is due and putting the remainder into a "pay over time" status, so you're racking up interest charges unintentionally. Or you may fail to notice recurring charges that you didn't intend to sign up for (political party solicitations often have "repeat monthly" pre-checked as a default choice, so you have to take care to uncheck them). If you don't review your credit card statements regularly, you may miss fraudulent charges, and there's often a time limit for challenges. Once it's passed, you'll have to eat the costs.

One major reason people don't save enough is that they can't picture how savings will turn into big numbers down the road. For the same reason, they can't appreciate how spending will suck up big numbers over time. I'm shocked to discover that someone might have $80,000 in a savings account and maintain $10,000 in debt, paying interest charges of 20% or more. They see that payment as a monthly obligation and never understand it as an aggregate.

It's very easy to turn a blind eye to how much debt you're accumulating through credit card obligations, and it's ridiculously easy to extend your payments. The credit card company is delighted to have you paying slowly as long as you do it regularly.

Virtual Transactions Have Actual Consequences

College students in particular get slammed with debt by making purchases using credit cards without a real understanding of the obligations. Young people who may never have met basic living expenses, and perhaps never even earned a dime, can't comprehend the enormity of the burden they are taking on. In addition to borrowing enough to pay huge tuition costs, they often must increase the loan to cover room, board, and other costs that increase the obligation substantially.

Studies show that 69% of 2018 college graduates took out loans and graduated with an average debt of $29.800. Some 14% of their parents took out an average of $35,600 in federal Parent PLUS loans. I'm told that the child of a family that earns up to as much as $250,000 qualifies for significant financial aid at one major private college at least, but many graduates are carrying huge burdens. There are 44.7 million Americans with student loan debt, and the total amount of US student loans is $1.56 trillion.[74]

With loan money in hand and a credit card in their pockets, many young people lead a lifestyle that isn't in line with their true circumstances. While it is true these days that even governments routinely spend above their means all the time without consequences, a newly minted college graduate has less chance to get off the hook. And unlike other obligations, even in the case of bankruptcy, college loans are never forgiven.

Most young people don't have a great grasp of numbers. They don't even seem capable of doing back-of-the-envelope estimates. An increasing reliance on electronic calculators is surely part of the problem. Even many college graduates admit to "innumeracy," the mathematical equivalent of illiteracy. In addition to an inability to quickly gauge the accuracy of a bill or calculate a tip, they have a shaky grasp of statistics and probability.

To add to this problem, young people who took out loans a decade ago have used forbearance to lower the monthly bite by stretching out the repayment period. The original rate was so high that despite making regular payments, some have seen their obligation swell to the point where they now owe as much as one-third more than the original loan. Worst of all, someone whose career goal doesn't materialize may not generate sufficient income to cover the loan payment as a budget item.

It's easy for students to rationalize taking on a big loan if they think of it as small monthly payments rather than focusing on the total amount. But the biggest danger of virtualized transactions like student loans is, of course, that the ramifications—because they're so far in the future—also seem virtual. As I have mentioned, we humans are wired for instant gratification as a matter of survival. When we were foragers and hunters, food was scarce, so when we stumbled on a bounty or made a successful kill, we gorged ourselves and added a layer of fat for lean times. But the tendency to overeat has negative consequences in a culture where food is plentiful. So does our tendency to open our wallets when we're feeling cash rich.

In the planning business, we use charts and graphs and thought-provoking icons to show our clients the dangers that may lie ahead, and we tell them anecdotes about people like them who behaved as they are behaving and what the results were.

Still: It's hard to change current behavior for consequences that aren't in your field of vision. And once again, technology can be used to negative effect. People who are invested in separating you from your money have become increasingly clever at it, as you may suspect, and research confirms.

Part Two

THREATS

4

Being Gamed

Letting Virtual Persuaders Push Your Buttons

Let's admit it. We in the consumer web industry are in the manipulation business. We build products meant to persuade people to do what we want them to do. We call these people 'users' and even if we don't say it aloud, we secretly wish every one of them would become addicted.

Nir Eyal, author and lecturer known as
"The Prophet of Habit-forming Technology"

THE IDEA OF MANIPULATING AN AUDIENCE didn't originate with the consumer web industry.

In 1657 the first advertisement for coffee appeared in a publication called *The Public Advisor.* Absent other means of persuasiveness, whoever wrote the copy relied solely on his imagination for all the virtues he attributed to caffeine to convince customers to buy this "very wholesome and physical drink." The writer claimed: "[It] *closes the orifices of the stomach, fortifies the heat within, helpeth digestion, quickeneth the spirits, maketh the heart lightsum, is good against eye-sores, coughs, or colds, rhumes, consumptions, head ache, dropsie, gout, scurvy, King's evil, and many others.*"

(I doubteth that.)

Two centuries later, advances in printing led to a proliferation of magazines and newspapers and, simultaneously, the invention of new tactics—some legitimate, some not—to attract customers. Quacks extolled the benefits of various "medical" potions, and in

1848 the tobacco company Cope Brothers & Co. was claiming that "smoke not only checks disease but preserves the lungs."[75] Abetted by technology, the manipulation business continues, with lots of new twists.

Deceptive Practices Updated for the Digital Age

All online businesses are focused on search optimization: ways to ensure that when anyone starts surfing the internet to look for the service or the product the business is offering, it appears at, or at least near, the top of the list. There are numerous ways to increase this likelihood. They include using keywords most likely to attract an audience, having a website that is well organized, and driving traffic effectively though social networks and other means. These are "white hat" techniques, so called to distinguish them from unscrupulous "black hat" practices. (The terms are derived from cowboy movies, where often you were able to tell the good guys from the bad by the color of their hats.)

One black hat technique is to use keywords deceptively. For example, a headline saying, "Angelina Jolie could use this for a houseful of cute kittens" might attract clicks from people interested in stories or photos about the actress and/or cats, but instead they'd find themselves at a website selling a vacuum cleaner good at sucking up pet hair. Or you're lured to a page that immediately redirects you to an ad or misleads you to click on an ad rather than the continuation of a story you're reading.

These techniques may generate more clicks for a website, although whether they generate more business is uncertain. In any case they don't do the kind of direct economic harm as duplicitous means that trick you into spending money such as these:

- *Charging for unintended purchases.* You order a phone and discover you've also been charged for a phone case that you never ordered. Or the seller obscured the fact that there would be extra charges (for delivery, for example). One airline ticket vendor used an online form that led you to purchase travel insurance that you may not have wanted. Those who resort to this sort of practice hope you'll fail to notice the additional charges, or that if and when you do, you'll absorb them rather go through the bother of protesting.

- *Forced continuity.* When you place an order, you're obliged to check a series of boxes that you don't realize commit you to automatic future orders. Or you sign up for a "free trial" for a newsletter or cosmetic product that extends automatically—for a fee—unless you are alert enough to uncheck the default box that authorizes it. Charities and political organizations often insert a "make this contribution recurrent" box that is checked as the default option. Time may pass and charges build up before you discover what's going on, and getting a refund may be a hassle.

- *Bait and switch.* One of the oldest tricks in the book, this now persists in digital form. You think you're taking one action but are actually taking another. You place an order and click on X—an action that normally closes down or eliminates a choice—but instead are ordering an upgrade.

- *Creating a false sense of potential scarcity and/or urgency.* You're considering replacing your golf shoes and find them online when up pops a warning: "There are only three left at this price." Or your search for a hotel for your vacation finds one "available for a limited time only" and/or you're advised that "booking now will save you 25%." You don't want to pass up a deal, but your impulsive purchase may cost you return postage or a cancellation fee.

Upselling and Upcharging

More problematic, because you are less likely to spot them, are the newer and more subtle ways that technology can chip away at your nest egg. In the last few years, for example, a taxi ride or a coffee shop or restaurant tab may have cost an additional 10% or more than it did previously—even if the cost of the product or service remained the same or didn't increase in the same ratio. Blame it on point-of-sale tap screens.

Just having the tap screen available at a group of fast-food restaurants increased the incidence of credit card tips from 28% to 60%.[76] The default tip options on the screens have also significantly increased the *size* of the average tip. In a New York cab, you can choose among 15%, 20%, or 25% tip options or a customizable number. Most people don't want to bother with calculations, so they make a percentage choice. And then, the combined effect of an "anchoring tendency" and the desire not to be seen either as an easy mark or a cheapskate leads most people to choose the middle number, 20%. Prior to the use of digital payments, tips averaged between 10% and 18%.

It turns out that even 20% might be considered too modest. Financial writers often note how a seemingly insignificant indulgence—a daily cup of specialty coffee—can add up over time to a substantial sum. And that was true even before a lot of retailers started using iPad tablets to process every transaction, no matter how small. After a woman at the coffee house register swiped my card for a $4 latte, she swiveled the screen toward me. "Add a tip," it said, with the suggested amounts of $1, $2, or $3—an upcharge of 25%, 50%, or 75%.

Social Media and the Rise of "Influencers"

With the rise of ad blockers and streaming platforms, the main way to attract attention for a product or service these days is through social media. Every marketer is looking for the most effective way to do it, and many turn to influencers. In fact, "influencers" have become a routine line item in a marketing budget.

An influencer is someone who can affect the purchasing decisions of others. It's a person who has become credible in one or more ways, including having demonstrated success or expertise in some area; having attracted other followers, which gives them authority; being attractive, which our innate attractiveness bias lets us assume they have other good qualities; and coming across as relatable, meaning "nice," "normal," or "someone I'd like to be friends with." An influencer has a following in a particular niche—the bigger the niche, the larger the following—and has influence over those followers.

While most influencers are real people, an unknown number of them may be computer-generated bots. A lovely, talented 19-year-old supermodel known as Lil Miquela who has 1.6 million followers is actually a bot that was created secretly in 2016 by an unidentified Los Angles company. Although 80,000 of her followers stream her songs, for two years most fans had no idea she was computer-generated until someone hacked into her Instagram account.[77]

Advertisers seeking an influencer to align with look for someone with an image and site content that's authentically compatible with their brand and message and, of course, with a reach and effect that can translate into sales. The online Instagram Marketing Hub offers advice for people wishing to become an influencer along with news, tools, and resources for connecting influencers, agencies, and platforms to one another.[78] It also offers an income calculator.

"A mid-level influencer," it says, "can charge $200 to $500 per post. It is not unheard of for those with more than 100,000 followers to earn $700 to $900 per photo. Those with 500,000 followers can command $2,000 to $3,000 per sponsored photo online." Kim Kardashian West, whose Instagram account in mid-2019 reportedly had 139.6 million followers, stated in a court filing that she commonly receives $300,000 to $500,000 for "just a single Instagram post endorsing another company's product that I like."[79]

Entertainers and sports stars, along with industry experts and others in the public eye, have long influenced popular taste and opinion and have been paid for doing so. But Huda Katten (27 million followers), Zach King (32 million), and Cameron Dallas (21 million) fit into none of those categories, and yet they made it into the top 26 on the 2018 Instagram Rich list, earning six-figure fees per post. And then there's Ryan, the 7-year-old star of Ryan Toys Review, whose $22 million earnings put him at #1 on the Forbes Highest Paid YouTubers in June 2018. They're among the growing number of "everyday people" who've become influencers, thanks to a gimmick and the drive to get attention by blogging, tweeting, posting pictures, and vlogging (posting videos, especially on YouTube, that are broadcast from a "studio" that might be their own living room or bedroom).

Social media helps them establish a more intimate connection than was previously possible in the entertainment and advertising worlds, and performers draw viewership rivaling (or exceeding) that of a conventional TV network show. Deals between viral performers and big brands are estimated to amount to $10 billion by 2020.[80]

What's different and noteworthy about sponsored videos is how hard they try to seem as if they're something other than what they really are—ads—because audiences today, especially younger ones, routinely tune out ads or skip them entirely. Since the FTC requires disclosure when content is paid for, the words "sponsored by" may show up, though almost imperceptibly, somewhere in the video: Even more subtly, they may incorporate an advisory that's very easy to miss: #ad.

Advertising That Aspires to Seem Authentic

Bloggers and others on social media also post #ad to indicate that they receive an incentive to promote the product or service they are writing about, but some now use #partner or #affiliate instead. To retain their "authenticity," they prefer to downplay their connection to a for-profit source.

With influencers commanding increasingly higher fees, it is not surprising to find advice from a marketing expert suggesting a less costly attention-getter to businesses and services—supplying free trial products and services to "perceived-high-reputation bloggers" whose opinion will influence shoppers. The expert also advises reaching out to "perceived-low-reputation-bloggers" to stress the usefulness and benefits of their products in hopes the bloggers will spread the word.

It doesn't seem especially cynical to assume that the prospect of getting a stream of free products (or better yet, paid advertising on their blogs) would predispose favorable comments from bloggers and/or a willingness to provide a "guest host" onsite space to promote a product. But bloggers who are successful have managed to gain their followers' confidence in their endorsements.

"It is impossible to overstate the influence of bloggers and product reviewers," says Robert Cialdini, an expert in the science of persuasion. "I've read that 98% of online shoppers read product reviews before deciding what to buy. That's a mind-blowing statistic. We can't get 98% of people to agree that the earth is round."[81]

Amazon is doubtless the most successful merchandiser in the history of the world. One measure of the power of its brand is that in 2015, 54% of people used Google for product searches and 46% used Amazon, but by 2018 those figures were reversed.[82] At Amazon, 90% of product views come from searches rather than advertising or other sources. The company has been uniquely effective at using social media for promotion. As you scroll down on any product page, your attention is caught by 12 different social value indicators that include recommendation widgets displaying everything from influencer-generated and user-generated content to average star ratings.[83]

The first online site to use ratings was eBay during the 1990s. At first, there were three ratings only: positive, neutral, and negative. Soon that system was replaced with stars. Now, star reviews are an entrenched part of many online sites—very prominently, and probably most effectively, on Amazon.

In 2016 Amazon vowed to crack down on "incentivized reviews," and its community guidelines say customers are now prohibited from "creating, modifying, or posting content in exchange for compensation of any kind (including free or discounted products) or on behalf of anyone else."

There is one exception: reviews of books and other items by the thousands of "trusted, established" reviewers who have been invited to participate in Amazon's Vine program based on how helpful other customers have judged their reviews. Identified as Vine Voices in their reviews, they can review anything, although a review of something from a Vine reviewer is labeled Vine Customer Review of Free Product.

Without reviews, products won't get clicks. And products that get just 3.5- or even 4-star ratings will need costly ad support. ("Neutral," as eBay discovered, is interpreted as "negative.") There's a paradox here. In order for a product to get clicks, people have to search for it. But it won't show up in searches if it gets too few clicks. With countless products available and multiple ways to buy the same product, vendors fight for attention. The three most obvious solutions are to advertise on Google (costly), hire influencers (results are uncertain), or participate in Vine.

When the Vine program began in 2000, a list of a dozen or fewer items, mostly books, would periodically appear online. Viners could vie for those they'd like to review, and they were limited to only a certain number of reviews. By 2011 the number of reviewers had grown tremendously, and there were as many as 4,000 items on the "reviewable" list. By 2016 there was one list targeted to a Viner's individual preferences and shopping habits and another list visible to all, with no limit to the products you could review.

If you invested $5,000 in Amazon in August of 2007 when Vine was announced, your stock would now be worth more than $100,000, John Herrman pointed out in a *New York Times* story. He added that if in the same time period you'd been a trustworthy and dedicated Viner, you might have accumulated stuff worth a comparable amount. He visited an active participant's home filled with boxes and goods—"five vacuums here, 14 hairdryers there, some laptops and cellphones . . . and pretty much anything with Bluetooth, so much Bluetooth, mouthful after mouthful of Blueteeth."[84] (Since 2015 the value of the products, calculated by Amazon, is deemed taxable, which induces many Viners to donate many goods to charity).

Amazon says it doesn't promise positive reviews by Vine participants or incentivize star reviews, influence the content of reviews, or even, in fact, require that a review be written. And it limits the total number of Vine reviews displayed for each product.

To vendors, Amazon notes that the process is particularly helpful for getting early reviewers for products too new to have generated high numbers of organic reviews. It also helps get traction for Amazon's own Basic Brands line and for products that are "exclusive" to Amazon made by private label companies, and it boosts Amazon Prime as well, because some items are available only to Prime customers. It has been reported that for participating in the Vine program Amazon charges $2,500 per item for its Amazon Retail partners (from whom it buys products wholesale and resells them).

In 2019 Amazon was testing a Vine program (with a much smaller cost per item) for its Amazon Marketplace partners, the much larger group that sells products on Amazon direct to customers and from whom Amazon collects a fee.[85]

Though they are increasingly using Amazon for product searches, people still consult Google, Yelp, and other review sites. To keep users coming, legitimate platforms have an incentive to keep their reviews credible and discourage any shady practices. By giving identifying information (such as their names, email addresses, websites) to a non-Vine source, which can be considered solicitation for free products, Viners may be dismissed as a reviewer. Yelp even has a form for reporting businesses that threaten to remove or manipulate reviews.

Of course, on a practical basis, monitoring is a big challenge for a large company. Although the number of vendors is hard to track down, in its 2017 annual report, Amazon reported two million marketplace sellers worldwide who represented 56% of the company's total business. (The latter number rose to 58% in the 2018 report.)

Using Data to Meet Your Needs and Tempt Your Desires

In the physical world, most of us routinely pass many storefronts. But unless something in the window lures us inside, we don't become customers. In the online world, however, merchants have learned how to take enormous sets of user-generated, user-specific data to know what we like and then put it right in front of us, convincing us to buy, or buy into, some thing or idea.

Not long ago, the owner of a local wine store announced that he had received a shipment of an especially delicious pinot noir. I got the news via an email he sent to me and presumably others whose affection for pinot noir he had learned of through data gathered online. And his effort had the desired effect: I did in fact consider dropping by the shop for a taste.

But that straightforward approach—identifying customers for a very specific product—is just the tip of the iceberg. What's really astonishing is how digital marketers have computers going to work as soon as you arrive on site—identifying you, calculating the products for which you're a potential customer, pooling you with others of a similar profile, and finally auctioning off ad space to marketers who want to reach customers with that customer profile. And the whole thing happens in a *nanosecond.*

With the right data, marketers can laser-focus on reaching a particular audience—say, exclusively women between the ages of 24 and 54 with a household income of more than $75,000 who spend a lot of time in the sun and regularly buy anti-aging products. Filtering is costly but ultimately cost-effective, since ads are directed only to targeted potential customers. A marketer selling men's underwear won't waste money advertising to females, and a toy manufacturer can direct his budget only to households with children. From random searches, sellers can make specific inferences. For example, once you've ordered green towels, manufacturers can send you ads featuring matching or complementary bathroom items.

Like other ad networks, Amazon (yes, Amazon again, because of its leadership role) uses technology to track customers from its site to other websites. But Amazon has an advantage over everyone else. Its experience handling so many millions of transactions, plus its role in delivering entertainment content, gives it unique and tremendous data-collection opportunities across a huge spectrum of categories.

Knowing that you bought a diet book and you're watching fitness videos lets it target you with an ad for a juicer. When you purchase knitting supplies and a book on grandparenting, you become a potential customer for portable cribs and baby toys. It doesn't give out your name, but it can help its clients direct their ads very precisely.

A real-life example: The ad agency for a company selling cheese bars, a low-carb snack, used algorithms to analyze how the company's search ads performed on Amazon. It discovered that customers who bought a lot of bars also bought a lot of trendy diet foods, including low-carb snacks and cauliflower pizza crust. They ran an ad targeting customers for those two specific product categories. When it was shown six million times over three months, it got nearly 22,000 clicks and more than 4,000 orders. That so-called "conversion rate"—the percentage of those who clicked and then went on to buy the product—was nearly 20%, a result the manufacturer called "amazing."[86]

Learning to Deliver Exactly What You Want

Economists Brent Neiman and Joseph Vavra from the University of Chicago Booth School of Business examined data on 160,000 households and more than 700 million purchases in grocery and big box stores for a decade. They found that despite many options, consumers are likely to stay with a single product in a specific category and unlikely to comparison shop.[87]

Fifteen years ago, for example, Tostitos was the dominant brand in tortilla chips, and it still is. But now there are 13 varieties, including lime-flavored, scoop-shaped, sea-salted, and multigrain. Since research has revealed our tendency to stay committed to our preferred lime-flavored Tostitos, the manufacturer knows it can be more inflexible about the price. If you want them, you'll pay for them.

Modern technology allows marketers to gather and process data that allows them to make broad generalizations like the preceding and also to discover and then appeal to our individual needs and desires. Delivering exactly what customers want is a powerful—and now widely used—marketing tool. We see it all the time: Once we've begun our vacation planning by searching hotels in Portland or flights to Mexico, we're not surprised to see pop-up ads promoting the glories of Oregon or assuring us that "Cancun Welcomes You." Or we browse for a gold-colored area rug and gold area rugs show up on every new screen we open.

In addition to what our own actions reveal online, marketers can gather or buy data that will tell them many other things: your age, ethnicity, marital status, and if you have kids; your alma mater, employment record, probable salary, and credit cards you carry; what section of town you live in, and when you bought (or lost) your home; your political, religious charitable leanings; what you read, what web sites you visit, and what things interest you. And they can use all this information not just to tempt you with goods but even to manipulate your thinking.

Using Psychological Tools to Shape Habits and Behavior

In 1957, marketing consultant James Vicary claimed success experimenting with an unprecedented marketing technique for which he coined the term "subliminal advertising." He used images flashed onscreen so fleetingly that they were unnoticed by most people—"perceived below the threshold of sensation or awareness." When he used it to promote Coke and popcorn during a film playing in a local theatre, Vicary said, sales shot up. Eventually, he confessed that he'd falsified his data, but the myth of "subliminal advertising" persists. While Vance Packard never used the term in his 1957 book *The Hidden Persuaders*, he did raise awareness of new marketing techniques that tried to manipulate customers through their hopes, fears, guilt, and sexuality.[88]

New findings in neuroscience and behavioral psychology have created an explosion of interest in what is now called "persuasive technology"—systems, apps, and software that through function or design can change attitudes and behavior.

Persuasive technology can serve us well in many ways, particularly by reinforcing good behavior. It might help in developing a health app that rewards you in some way as you progress toward a fitness goal. No doubt it helped inspire the Progressive Insurance program that monitors your driving data and rewards safer driving with lowered premiums.

But to make us more susceptible to their messages many, if not all, major companies use some form of persuasive technology to serve their own ends. And it's perfectly legal.

For instance, on the Pinterest site the bottoms of images are always cut off to encourage visitors to keep scrolling down. At the end of one You Tube video of *America's Got Talent*, another cues up almost immediately and then another and another, basically hijacking your attention. Snapchat motivates engagement by distributing badges to users who maintain steady streaks of communication, posts an hourglass emoji when the streak seems to be coming to an end, and awards a "100" badge if a streak continues for 100 days. This is all based on a well-studied psychological need to bank progress. Facebook uses persuasive technology to determine what appears in our news feed.

GSN, a company that designs online casino games, uses data it pulls from phones to measure the speed, frequency, and enthusiasm of a player pressing the screen. With that information, it can predict when the player is about to lose interest and will then immediately suggest other games. While the games are free, the company sells ads and virtual items of interest to players; the longer players are online, the better their bottom line.

The staff of an advertising agency today is very different from the one you might have seen on the TV series *Mad Men*, which takes place in the 60s and where the creative team consisted of wordsmiths and artists. Today, the team is likely to include people who hold PhDs in game theory and predictive modeling, a system of statistical analysis to predict outcomes. Using artificial intelligence (AI) and collected data, the team can tell clients which particular ad to show a given customer on a specific website.

All big retailers, from the postal service to the supermarket chains and discount stores, have "predictive analysis" departments to figure out buyers' personal and shopping habits. Their work is based on the science of habit formation, which for a couple of decades has been the subject of research in neurology and psychology departments at universities, medical centers, and corporate laboratories, and is the subject of Charles Duhigg's revealing *New York Times* story in 2012, "How Marketers Learn Your Secrets."[89]

Duhigg points out that 45% of the daily decisions we make may be based on habit rather than conscious choice. To conserve effort, our brains try to convert regular behaviors (from simple ones, like brushing your teeth, to more complex, like backing out of the driveway) into an automatic routine—a habit—through a process called "chunking."

It's a three-step process: a cue (that triggers your brain to start the routine), the routine itself (physical, mental or emotional), and the reward (which may be the simple satisfaction of getting the job done). The process becomes completely automatic, and it will be repeated unless you deliberately try to change any part of it.

Even where you shop and what you buy at various stores (like those lime-flavored Tostitos) turns out to be habitual behavior that is very hard to break—except at times of certain life events. Among these events are getting married, moving to a new home, and having a baby.

Target, the retailer, was especially interested in developing its customer base by changing the habits of expectant parents, people who (like many customers) visited Target primarily to shop for cleaning products. Of course, once the baby came, the parents would not only be shopping for new types of items but also, with their newly expanded households, would be potential long-term customers across many categories.

Many retailers routinely check out birth records and start to woo new parents, but Target wanted to start attracting these customers before the baby arrived. Mining their data, they found early clues that a woman customer might be pregnant. For instance, she might switch to unscented lotions from her regular brand, add supplements like calcium, magnesium, and zinc, and toward the end of the pregnancy start purchasing supersized bags of cotton balls and other baby supplies.

The company already had the capacity to send every customer a specially designed ad book. So mailings to a woman whose purchases suggested she might be pregnant began to include baby-related items (carefully mixed with unrelated items to seem random enough that the woman didn't feel her privacy was compromised). The goal was to change her habits and persuade her to consider Target not only as a source of cleaning supplies and drugstore items but also as a resource for baby items and, over time, other goods.

Exploiting What Motivates You

It's not just your buying preferences and interests that are being tracked. Motivated parties are trying to figure out something more subtle and potentially creepy: what motivates you. They can reverse engineer that information.

"Future applications making use of big data, location, maps, tracking of a browser's interests, and data streams coming from mobile and wearable devices, promise to usher in the era of unprecedented power in the hands of marketers, who are no longer merely appealing to our innate desires, but programming our behaviors," Bill Davidow wrote in *The Atlantic* in 2013.[90]

Marketers, he suggested, are borrowing ideas from B.F. Skinner, the scientist who put pigeons and rats in boxes and performed experiments to modify their behavior through punishments and rewards. Our pecking on our cell phone keys is similar to pigeons' pecking on buttons for food, he says. Our behavior has become a three-step habitual process: the cue (cell phone ding!), the routine (of answering), and the reward (blog post, email, cat video).

"The beauty of the internet is that by combining big data, behavioral targeting, wearable and mobile devices, and GPS, application developers can design more effective operant conditioning environments," Davidow wrote. And each of us, with a smartphone in our pocket or handbag, is in a virtual Skinner box: We hear the ding, we answer, and we get the reward.

Eventually, a fast-food provider monitoring your data may be aware that you are trying to lose weight and knows that you might be susceptible to impulsive cravings for sweet and/or fatty food. If you're near a franchise, that marketer might be able to send you a discount for a shake or fries—a classic cue/routine/reward, which in this case is a prompt to take an action that is contrary to your actual intentions.

The stakes might be much higher. As virtual experiences become more real and behavioral targeting more precise, marketers will doubtless find expanding opportunities. If you can "try on" outfits virtually and show them to your friends on social media, their compliments would tempt you to buy. (As we know, social media tends to increase your spending habits.) Someone who has a thrilling virtual driving experience at the wheel of a supercharged sports car is likely to take up an offer to test-drive an actual vehicle as he's passing a dealership.

A Perfect Storm of Temptations Affects Your Bottom Line

While the influence of social media ramps up your spending, an even more pervasive way technology can impact your financial future is by having made the very act of spending so much easier.

When I talk with older people who are contemplating a downsizing move or to a client faced with the task of cleaning out a deceased parent's home, they tell me how overwhelmed they are by the amount of stuff they have to clean out, much of it consisting of purchases still in the box, items that tickled a fancy but were soon recognized as follies: a rechargeable jumper cable that no one remembered to periodically recharge, a juicer too bulky to leave on the counter and too much of a nuisance to clean, the camera left to gather dust when you relied on the one in your smartphone.

In fact, a really useful app might be one called "Seriously? Do You Really Need That Thing You Just Put in the Cart?" whenever you tried to access PayPal or the Amazon payment authorization page. That would be a technology expense worth paying for.

How crazy is it that one of the best-selling books of the moment—Marie Kondo's *The Life-Changing Magic of Tidying Up*—is about getting rid of stuff, and that huge industries have been spawned to deal with the results of our having bought too much.

The company 1-800-Got-Junk hauls away and disposes of items you once thought you needed. (It's testimony to how desperate people are for these services that in some places, they'll even show up at your home until midnight.) By 2019 the company reported that it was operating in three countries with 160 franchises and had generated $182,500,000 in revenue the previous year.

The self-storage industry is booming. In the past, once we ran out of room to put stuff in, we stopped buying more stuff. Now, we just pay for more space to put it in. By 2019, storage unit rental had become a standard budget item for 9.4% of US households. It costs an average of about $1,200 per year to rent a climate-controlled 10 × 10 room to store stuff that you visit maybe once or twice a year—very likely on occasions when you come to add more.[91]

It's an ongoing source of amusement in our company that when a particular colleague is laid up at home for a couple of days, packages from Amazon addressed to him will soon start arriving at our office. With time on his hands, he succumbs to that popular pastime, internet shopping.

Online purchasing is certainly efficient. In the past, if you were hunting for a particular record album or pair of shoes, you had to personally visit stores. There was a self-imposed limit to the amount of time and patience you had to spend. Remember the ad for the now-defunct Yellow Pages, "Let Your Fingers Do the Walking"? On the internet today, your fingers can run marathons. According to scrapehero.com (a company that extracts data from websites), by January 2018, Amazon alone offered 564 million products for sale in the United States—an increase of 208 million from the previous year—and offered a total of three billion products across its various marketplaces in North America, Europe, and Asia.

And now that more and more users are accessing the web from a tablet or smartphone rather than a computer, it's possible to make impulse purchases just about anywhere and at any time. During the Thanksgiving-through-Cyber-Monday peak shopping period of 2018, nearly 56% of online sales came in on mobile devices. It's even easier to simply call out, "Alexa, buy me some olive oil," as you chop vegetables on the kitchen counter. Whatever you forgot to put on the list or you have run out of, Amazon can add to your cart.

Furthermore, Amazon now holds patents that could instruct its Echo device to listen for certain trigger words (like "vacation plans") in a user's recorded conversation that it could transcribe and direct back at the user in some version as a selling tool. It's not being done yet, says a company spokesperson, adding that "patents take multiple years to receive" to downplay concerns.[92] (Yet, with the increasing connectivity of Internet of Things devices, paranoia is inevitable. I actually had a call from a friend who started talking in a whisper because he didn't want Alexa to overhear him talking about a particular restaurant and take that as a request to call for a reservation.)

When you ask Alexa to buy something, unless you specify otherwise she orders from the Amazon family. Although the company started out to democratize selling—allowing small retailers to peddle their wares alongside the behemoth operation—it began selling its own products in 2009, initially basic items like power cords and batteries. In the latter case, its prices were as much as 30% lower than national brands, and soon Amazon had one-third of the national online battery market.

As the earlier example of lime-flavored Tostitos demonstrated, the more specific a niche a manufacturer's product occupies, the more price control the marketer can impose. As long as there are lots of battery options, one company can't control the market.

At a 2018 conference, Amazon introduced an astounding 70 new products, including a gadget to bring Alexa into your car and a smart plug that automatically connects to Wi-Fi and makes any appliance automatically respond to your commands.

Amy Webb, author and futurist, says Amazon, Google, and Apple are all using AI to transform our homes into data-generating machines to track and "productize" (there's a new word that you'll probably be seeing a lot more of) daily activities.[93] They are all offering smart speakers and partnering with appliance makers, and one or more of them is getting into the credit card business, home alarm systems, and health care.

Once your data, gadgets, appliances, cars, and services are entangled, you'll be locked into Amazon, Google, or Apple. As you buy more stuff, Webb warns, you'll find that not only are your devices incompatible with other systems, but also you and your family will be effectively segregated by the brands you use.

Apple's products are the technologically simplest, but also the priciest, and will be bought by those who can afford them. Currently egalitarian Google could shift to a future of tiered services and permissions. The currently small clusters of Amazon homes with smart speakers, door locks and e-keys, video doorbells, and thermostats might be the start of entire developments full of appliances connected to an Amazon platform. It's a potential caste system in the making.

Technology doesn't *force* you to buy things, but it's very effective at tempting you. And anyone with an internet connection has access to a cornucopia of goods that far exceeds whatever was available to the richest man in history—who, for the record, was King Mansa Musa I (1280–1337), ruler of a West African Empire. His inflation-adjusted personal net worth was $400 billion, eclipsing the $150 billion fortune belonging to the world's richest man, Jeff Bezos, in January 2019 (before his divorce).

As for Mansa Musa I and his family, when they failed to fend off civil war and invading conquerors, his holdings disappeared within two generations. The powers of technology could blow a giant hole through yours in far less time.

NOTE: The world of technology is constantly evolving. The following information is accurate as we go to press, but some of the sites may have changed names, become irrelevant or obsolete, disappeared, or in some other way become unreliable. It's my hope that you'll get some good ideas and use what follows as the jumping-off point to do your own research.

Strategies for Online Shopping

Resist These Upsell Ploys

- *Free shipping/discounts on purchase*: To get these offers, you may be pushed to a minimum spending level.
- *Easy checkout*: You're already on file, so it's easy to impulse buy with a click.
- *Recommended list*: Unnecessary items that complement your purchase or you may have browsed in the past.
- *Reminders you've left something in your shopping cart*: "Just browsing" to kill time, you may have been momentarily tempted by something you don't really need.
- *Anchor Pricing*: Before you purchase the so-called better value of two (often high-end) items placed side by side, do some research.

Always Comparison Shop

- The more specialized the demand, the more control a company has over pricing.
- Clear cookies and check prices online before you log in, to see if you'd be offered a better deal as a new customer rather than as a returning customer.

- Be aware that prices from a single seller may vary depending on multiple factors. Amazon started the practice of dynamic pricing, according to its website, in order to meet or beat the lowest competitive price. Competitors have tried this, sometimes charging different amounts online and in stores, or changing prices in different locations (now against the law in some states). Jet.com (a Walmart company) has been offering a "real-time savings engine" that drops prices as consumers buy more goods. The trend is for pricing to become more localized, more dynamic, and more personalized. Some predict that in five to ten years, everything you buy will be based on more personalized numbers. That will be much easier, the more cashless our society becomes, but it may not benefit consumers.

Take Precautions When Ordering Online

- Verify that the retailer is using a secured connection by making sure that its Web address starts with the "https" prefix (note the "s" following "http" means your info is encrypted) and that there is a locked padlock icon on your browser's status bar.
- Avoid sending financial or personal information via hot spots or on public computers or an unprotected smartphone.
- Before shopping with an unfamiliar online retailer, check it out at the Better Business Bureau. Confirm the address. Search the company's name together with words like "reviews" and "complaints"
- Use a credit card rather than a debit card; it's easier to dispute charges.

- Read the information regarding delivery times as well as backorder and return policies. (Is there a time limit on returns? Can you save shipping charges by returning items to a brick-and-mortar store? Will you be charged a restocking fee for a return?)

Be on Guard for Fake Reviews

I've been told that 84% of people can't spot a fake review. Here are some clues that should raise your suspicions:

- *Reviews clustered around the same period.* If a product has launched or started a campaign, they may be soliciting reviews. (Of course, around gift-giving holidays, there may be more people getting new products and posting more reviews.) If you sort from newest to oldest, you may find initial positive reviews sent in as reputation builders, but over time the reviews may be more accurate.

- *Lack of details.* Authentic reviewers give very specific information and mention specific items. Hotel room reviewers might talk about "check-in" or "distance from elevator"; restaurant reviewers might mention the "noise level" or describe specific dishes.

- *Repeated words.* When you spot similar phrases in multiple reviews, it's likely the reviewers have been coached.

- *Lack of photos and/or reviewers with generic names*: These reviews are more likely to come from a paid source rather than actual customers.

- *Poor English*: Such reviews may have been outsourced to international content "farms."

Use Online Help to Check Review Authenticity

- Fakespot, a free site, analyzes product reviews from various companies. Copy and paste the link to the product page, then click Analyze. You can also use a browser extension for Chrome, Firefox and Safari: click the Fakespot icon in your toolbar. Also available for Android and iOS. Gives reviews a letter grade.
- ReviewMeta, Amazon-specific, strips out or reduces the weight of some reviews to give you an adjusted rating—that is, what the Amazon average rating number would be if the questionable reviews didn't exist.
- Bottom line: None of these approaches is absolutely foolproof, and everything you get on the internet is always in flux. Check things out as best as you can, and use your common sense.

5

Giving Up Privacy
When They Get to Know You, They Know How to Get You

I don't know why people are so keen to put the details of their private life in public; they forget that invisibility is a superpower.

Banksy, street artist and disrupter

"If you're not paying for it, you're not the customer; you're the product." The phrase dates back to early television days, but it seems particularly relevant today. Advertising is of course meant to attract eyes and ears, but companies like Facebook know enough about you to direct their efforts specifically toward *your* eyes and ears. When Facebook sells an ad targeting someone with your particular browsing history, you are being sold in a way that is quite different from being exposed to an ad on Monday Night Football.[94]

Whenever you're looking at "free" content on the internet—a friend's Facebook post, a piece of luggage on eBay, a live news feed, or a video of cockatoos that dance to Elvis—you're relinquishing something in return, and that happens even if you've paid a fee. Once you've clicked on any host site, it's almost certain you're giving away data—information that can be of tremendous value financially or ideologically to any number of parties.

The idea of collecting data for marketing purposes is a century old. Arthur Neilson founded his media research firm in the 1920s, expanded it into radio market analysis, and then applied his skills to television. The information he gathered on audience size and makeup (age, gender, race, economic class) has long been used to establish ad rates.

From its inception in the late 90s, the search engine Google kept data on click patterns and locations of users, although neither Google nor Facebook adopted a free model in order to serve advertisers. In fact, they adopted the advertising model so they could keep serving their users for free. But they haven't taken care to see that users understood exactly what data they were giving up or how it might be used.

The game-changer was the arrival of sophisticated website analytic and tracking software that made it possible for advertising to be targeted and relevant in an unprecedented way. As the technology has become more sophisticated, it has become capable of collecting increasingly detailed information on everyone who visits a site, information that can be utilized—some say *weaponized*—to motivate us in multiple ways, from making a buying decision to subscribing to a belief.

Every one of the billions of internet users is a data-producing machine, and users become metrics. Site owners soon recognized that they could use the data they mined not only for their own purposes in new and innovative ways, such as targeted messaging, but also as a commodity they could sell to others who had their own products and messages to promote. Content attracted users and users provided data. For motivated parties—purveyors and persuaders, marketers and manipulators—data was the tool they needed to influence potential customers, followers, or both.

Facebook focused its efforts during the years 2012 to 2017 on "growth hacking"—a term coined to describe every startup's make-or-break need to grow or else to die. Growing the company is the only thing a growth hacker does, but growth hackers may have various kinds of expertise, from marketing to engineering, to develop everything from design tweaks to improved algorithms. Facebook became an unparalleled success at collecting data and monetizing the results.

If You Give a Marketer a Cookie . . .

The children's book *If You Give a Mouse a Cookie* is a cautionary tale. Once a hungry mouse shows up on your doorstep and you give him a cookie, he'll want milk. Then he'll want a mirror to check if he has a milk moustache, and the demands continue. There's a parallel to this in the data collection business.

When you visit a website, it can tell your browser to create a so-called "cookie" that holds data about you that tracks your activity and preferences. *Session cookies* are small files that make using the website easy. For example, a session cookie tells it to communicate with you in English and to allow you to move from page to page without repeatedly logging in. The session cookie will expire when you leave the site or shortly afterward.

Persistent cookies, sometimes called *stored* or *tracker cookies,* are stored on your computer or other devices by websites that get repeat visitors—such as eBay, Amazon, and even Gmail. When you return, the tracker cookies act like an ID card. They remember your name, password, language preference, and even perhaps your credit card information (encrypted for your safety). This type of cookie remembers what city you're inquiring about when you visit a weather website. It also keeps the stuff you tossed into your cart when you were insomnia browsing, hoping to tempt you to "Check Out Now" when you return. These cookies expire eventually, but not as quickly as a session cookie.

Companies also use persistent cookies to track your behavior—like how often you logged in and what you looked at—in order to improve their product or send targeted ads. Shopping sites often request that you log in before you browse so they can build their database and send you ads.

Even sites you don't log into (like news sites and blogs) may track you. Although they don't know your name or email, they can give you a user ID and collect information for various reasons—perhaps to improve their site or even to affect the pricing (charging more if you live in a particular neighborhood, less if you live near a competitor's store, although some states disallow location pricing).

Third party cookies, also called *web-serving ad trackers*, are different. They are collected by advertisers, not by the website. They collect data on a user (age, location, interests, etc.) and the user's behavior (what he or she has viewed or clicked on). Single-website trackers manage data for individual clients. Multi-website trackers share users' activity among multiple clients and are more invasive. They can combine data to get a more complete picture of you and then show you more targeted ads with prices correlated to how much you are able to spend, and they even alter content based on what they think you want to see. Even more insidiously, they are more likely to have the capacity to uncover your real-world identity.

The Exponential Growth of Data Collection

Visit http://webkay.robinlinus.com to see the data your browser can reveal to any website without your granting permission. Seconds after arrival at a site, the owner can easily determine your IP address (your computer's location) with 99% accuracy by country, 90% by state or region. It can identify with 81% accuracy your city's location within 25 miles in the US, and outside the US or Europe your city location within 25 miles can be determined with 60% accuracy worldwide. It is also possible for the site owner to identify the browser you are using and other information about you.

Retailers can discover what keywords led you to their site, what in their advertising appealed to you, what interests you and what gets your attention. Data extracted while you search for vacation information will help merchants determine how best to market their airlines and hotels to you. You can be tracked as you compare two lamps on a single website and then followed to another website where you eventually purchase a third.

Collected data may even determine the price you will be offered—not only by location, but also with the aid of a retailing process called dynamic pricing that I also mentioned previously. It's based on supply and demand. An airline ticket price may be established depending on the type and number of seats available and the time remaining until the flight departs.

Also thanks to data collection, your past Google searches can come back like zombies in the form of "stalker ads." If you briefly looked for sneakers or even left a pair unbought in a shopping cart, these ads may keep popping up, even sometimes long after you've bought them.

Collected "data" can also be much more personal: your age, gender, income, diet, weight, browsing habits, allergies, and job title—and much, much more. Facebook may suggest that an advertiser target you based on information you would never expect to emerge, much less be used to influence you. For instance, data about you may reveal that you have become engaged or that you have a close friend who loves soccer. Every platform is trying to collect such detailed data to provide to its advertiser clients.

Furthermore, *The Wall Street Journal* reported that intimate data women recorded on several health care apps was shared with Facebook without the women's knowledge or consent. The information included the user's weight and height, plus her selection from the menu of problem areas to work on, such as "belly fat"; it included facts about a user ovulating or menstruating and engaging in 46 minutes of sexual activity, information a woman provided (in case you're wondering) in order to get a calculation of calories she had burned that day. (After the story appeared in the newspaper, some of the apps changed their sharing policies.)[95]

In 2016, Facebook was sued by a patient with metastatic cancer for violating his privacy by collecting data about his participation on cancer websites outside of Facebook. The case was dismissed and at last report was under appeal.

As more people have become aware of cookies and made a habit of deleting them or downloaded an ad blocker to do the job automatically, marketers have had to look for another route to find their audience. So they have turned to device fingerprinting to identify customers. By tracking various characteristics—location and time zone settings, the operating system, installed apps and plug-ins and browser versions, etc.—device fingerprinting makes it possible to profile a particular device and connect your online identities to one another.

Some 70% of people log onto the internet regularly with at least two devices (most commonly a phone and a laptop). Device fingerprinting makes it possible for interested marketers to send a targeted message to both of them as well as to any other device (like a tablet) their owner may use. Suppose you click on an ad on your phone for camping gear and subsequently do a search for the gear on your home computer. If you place an order online, fingerprinting that identifies your phone would let the store know you used the browser on a computer with the same fingerprint as the device where you saw the ad—so they'll know which ad led you to buy the gear.

The Internet of Things—the increasing interconnectivity of all appliances by chips—compromises our privacy and leaves us vulnerable to an even greater degree. Social media operators, tech companies, and app makers want to collect as much data about you as possible online—where you are, what you like, how you date, what sites attract you. By some estimates, 30 billion interconnected devices are currently in use, and the process is already underway.

It is possible to remotely activate the camera on your phone and the microphone on "virtual assistants" to record conversations and even take videos unbeknownst to us. An Amazon device recorded a woman's private conversation in her Portland home and shared it with one of her husband's employees in Seattle. Amazon later confirmed that Alexa mistakenly heard a series of commands and sent the recording as a voice message. Like other electronic assistants, Alexa as of early 2019 wasn't authenticating or authorizing individuals who access it, so it could be directed by anyone familiar with the owner's commands. (Some electronic assistants offer means to enable layers of privacy protection, but that requires that you read—and understand—the manual.)

What's more, the voice recognition technology is just getting better. Old NASA research on subvocal speech recognition is being explored to find "non-acoustic modalities of subvocal or silent speech recognition." That means if for privacy reasons you choose not to say things aloud, they're trying to find methods of recording surface electromyographic (sEMG) signals "from muscles of the face and neck" which, tech writer Alistair Dobbs suggests, "may make it possible for you to accidentally tell next-generation smart devices to buy 400 rolls of toilet paper by simply clearing your throat. Just what no one always wanted."[96]

Cable and satellite providers are already customizing TV ads. At the same viewing time a household with a cat now gets commercials different from those shown the neighboring household with a dog, based on information from credit card data, CPS phone location data, and other data.[97]

A company called Samba TV pays a fee to the manufacturers of smart (internet-connected) TVs to embed its technology in their set. The manufacturers get extra income when TV purchasers accept an offer (as 90% do) to get the service Samba provides: recommending shows based on your past viewing selections.

(By mid-2018, Samba had already collected data from 13.5 million television owners. In 2019 Twitter entered a partnership with Samba to help measure Twitter's effectiveness in driving tune-ins to primetime broadcast and cable TV programming, streaming originals, and sporting events.)

Samba can track what someone watches on a second-by-second basis and offers advertisers the ability to base their targeting on whether a viewer is watching conservative or liberal news outlets.

Even better, Samba can identify and send targeted advertising to other devices that share the TV's internet connection, like tablets and smartphones, so the company can track users "in their office, in line at the food truck, and on the road."

In other words, you're not only being targeted while you're channel surfing from your sofa. Unless you turn off the location tracker on your cell phone (and many people aren't even aware it's *on*) it tracks you everywhere you go, and it goes so far as to identify the specific stores and offices you visit. A firm or retail outlet targeting new clients can hire people to harvest such data, using "geofencing" to track persons in specific counties who have visited one or more competing firms, then put its own ad on that person's phone.

The interconnectivity of devices, together with the technological capability of zeroing in on who is operating a device, allows unprecedented intrusion into your space, both private and public.

In fall 2017 a giant billboard screen went up in London's Piccadilly Circus that used built-in cameras to analyze the make, model, and color of approaching cars to get a profile of the driver and then flash customized advertisements onscreen. We're going to see a lot more of this sort of thing. Because mobile phones have replaced TV as the most valuable source of consumer data, US ad dollars spent on digital billboards grew to $8 billion in 2018, up 5% from the previous year.

US Government Oversight Is Just Beginning

Until recently, our elected officials haven't responded to privacy concerns. In fact, in 2017 Congress voted to repeal the FCC's broadband privacy protections that had barred internet providers like AT&T, Comcast, Time Warner, Sprint, and Verizon from collecting, using, and selling personal data like web browsing history, email addresses, and first and last names without prior consent.

The EU took an opposite stance. It established the General Data Protection Regulations (GDPR) that went into effect in May 2018 to set some limits on how much information could be gathered and under what circumstances.

The rules granted internet users the right to ask how their personal data is collected, stored, and used and also the right to ask that it be deleted. Under the rules, protected data includes your name, email, and IP (device) address as well as information under a pseudonym that could be tracked back to you. You can even prevent your data being used for a special purpose. For example, if after you buy a messenger bag online you begin to see ads for messenger bags appearing on your devices, you can ask the retailer to stop using your personal data for direct marketing.

The rules also stipulate that companies must have some kind of legal basis—such as actual consent—to collect personal data. (And in March 2019, the Dutch Data Protection Agency went further. It said that websites could not refuse access to visitors who wouldn't agree to be tracked.)

Without consent to collect data, companies risk fines up to 4% of their global revenue or 20 million euros (whichever is higher). That's a large enough number to be of consequence: On Apple's 2017 revenues of $229.2 billion, the fine would have amounted to $9 billion. And the legislation does affect Apple, since it applies not only to the EU but also to any US companies to which the following applies:

- The business markets to or collects information from customers who are EU citizens.
- The business is a data collector and processor (including cloud service providers, accounting services, payroll companies, data disposal services, IT service providers, and payment processors) that has frequent contact with personal user information.
- The business collects "personal" or "sensitive" information. The former applies to any data that can identify an individual and the latter consists of information that includes the following:

    Racial or ethnic origins

    Political opinions

Religious or philosophical beliefs

Health information

Sex life and sexual orientation.

California became the first state to act on its own to protect consumer privacy with the 2018 Consumer Privacy Act, effective as of 2020. It gives consumers broad rights to take back control of their personal information. Vermont, Colorado, and New Jersey have new privacy enforcements, and some are anticipated in New York and Washington.

In spring 2019, legislation was introduced in Congress to strengthen the privacy rights of children. And rather than optional Do Not Track settings (available in most browsers but generally ineffective), Apple is going all in with Intelligent Tracking Prevention in the newest version of its Safari browser. It's focusing on privacy by allowing long-term persistent cookies and website data only from sites users actually interact with and removing other tracking data proactively.

There were news reports in June and December of 2018 that Facebook had given business partners—including makers of smartphones, tablets, and other devices—deep access to users' personal information, letting some companies effectively override users' privacy settings. Then came news that the Department of Justice was pursuing a criminal investigation into Facebook, which had cut deals that let more than 150 companies, including Apple, Amazon, Microsoft, and Sony, see Facebook users' friends, contact information, and other data, sometimes without consent. By the spring of 2019, Mark Zuckerberg was speaking publicly about the need to protect privacy and calling for more regulations. (Having regulations to abide by would, of course, make Facebook less vulnerable to lawsuits.)

Our privacy is compromised in many ways—but our own complicity is part of the problem.

"Privacy Is Dead, and Social Media Holds the Smoking Gun"

The quote is from Pete Cashmore, the CEO and founder of the digital media website Mashable.

Anyone with a social media account is likely to have posted what amounts to a bonanza of identifying material: names, email addresses, and locations; employment history; names of family members, friends, and pets; birthday and anniversary dates; and other clues that help motivated people crack passwords or answer security questions.

The point was made in dramatic fashion in a television ad from Febelfkin, the nonprofit Belgian Financial Sector Federation. A series of passersby (real people, not actors) are randomly invited into a tent to have a session with a psychic, and each in turn is stunned as the psychic comes up with incredibly specific insights: He describes the color of one man's motorcycle, reels off the exact amount another woman spent on clothing the previous month, and even reveals that a third person participated in a sexual foursome. "Hardly anyone knew!" she gasps.

At the end of the sessions, a curtain is dramatically pulled back, revealing a small group of researchers in a corner of the tent. Facing a bank of computers as the psychic subtly pried clues out of the subjects, they had quickly gathered more details about the subjects online and transmitted them into the psychic's hidden earpiece. You quickly realize that the subject themselves had publicly revealed all the information that the psychic "intuited."

We say we are concerned about protecting our privacy. A 2015 Pew Research study found that 74% of respondents said it is very important to them to be in control of who can get information, 61% said they would like to do more to protect their privacy, and a large majority said the privacy laws should be tightened and advertising should be regulated.

But there's a paradox at work. Though the Belgian commercial is just one example of how much data can be harvested from social media postings, not everyone is willing to stay away.

While 59% of users thought giving up social media would not be hard (of whom 29% percent said it wouldn't be hard at all), 40% held the opposite point of view (with 14% saying it would be very hard.) As you might expect, older people were less engaged with social media. Among people 50 and older, 66% would find it easy to give up vs. 33% who would find it hard. For people aged 18 to 24, the respective numbers were 49% vs. 50%.[98] In fact, being tracked, rather than being seen as a problem, may actually confer a sense of "specialness" to the younger demographic, as three studies of Harvard students suggest:

- When offered a discount coupon to a restaurant online, students were more likely to buy it if they thought they were targeted for the offer based on the travel sites they had previously visited rather than randomly or on the basis of age and gender.
- Shown an ad for an expensive watch, some subjects were told the ad was targeted to them individually based on their browsing habits, and some were told the ads were untargeted. Asked later to rate how sophisticated they considered themselves to be, those shown the targeted ad rated themselves more highly than the others because they thought being targeted was a reflection of their good taste, and this also raised their interest in buying the product.
- In a third study, after receiving what they were told was a behaviorally targeted ad for an environmentally friendly product, students rated themselves as more "green" than others and were more willing to purchase a green product and donate to a pro-environmental charity.[99]

We Sign Away the Right to Protect Our Privacy

As far back as 1998, the FTC had strongly suggested that websites disclose what they do with the information they gather and what choices you as a visitor have in that regard. Nevertheless, many companies ask or even require that customers forfeit a very comprehensive list of rights by signing a Terms of Service Agreement (TOSA), which includes the Privacy Policy—and they're met with little to no resistance.

Computer users in the United States received a flurry of e-mailed privacy notices and consent forms in early 2018 that were a direct result of the new EU privacy laws. Most of us continued to do—or rather not do—as we had already done. We signed the service agreements without reading them. There are several reasons for this.

Reason #1: We need access to many sites. We use them to work, to get places, to save time, to communicate.

Reason #2: We (mistakenly) think our privacy is protected. Surveys vary, but anywhere from 54% to 73% of people assume that simply having a "privacy policy" means a website won't share your information with other sites or companies without your permission. *Nothing could be further from the truth.*

Look, for example, at a recent version of Target's TOSA and especially its section on privacy policies.

To start with, it seems *designed* not to be read. Type is small, nothing's in bold, and some sections are all in caps—which makes them harder to read. Then, it obliges a customer to agree to let them collect name, address, date of birth, driver's license and credit card number and much more, including:

- Mobile device information (e.g., device model, operating system version, device date and time, unique device identifiers, mobile network information)

- How you use its sites and mobile applications, search terms, pages you visit on its mobile applications and application performance
- Geo-location and in-store location
- Social media information (e.g., ID, profile picture, gender, age range, friends' IDs)

If by (the very unlikely) chance you did read more closely, you'd see that you've given it the right to use information it collects for its own purposes—and also to share it with third parties. These are some of the other conditions imposed by the Target TOSA:

- "We collect data that is publicly available. For example, information you submit in a public forum (e.g., a blog, chat room, or social network) can be read, collected or used by us and others, and could be used to personalize your experience."
- "If you choose not to provide information, we may not be able to provide you with requested products, services or information."
- "Target reserves the right to make changes at any time."
- "Each time you use the Site, you should review the current Terms & Conditions."

I guarantee you that if there are changes in the current Terms & Conditions, they matter—not to you, perhaps, but to the company. The changes are there because some unforeseen problem arose that made the hairs stand up on the necks of the company attorneys, who immediately added more words designed to protect the company if it ever happened again. What you really need to know is the real-world reason for the changes. That's not likely to happen. Neither is it likely that you will review the current Terms & Condition regularly.

Reason #3: We don't have time to read all those policies. A Carnegie Mellon study from 2012 found that the mean length of the TOSA agreements of the 75 most popular websites of the time was 2,500 words. Doing an online study of 212 participants, it calculated their average reading speed and the average number of sites (106) they visited per year and concluded that they would have to spend 244 hours a year reading TOSA agreements.[100]

By now, I would think the average person visits more than 106 sites per year and (if the attorneys have been doing their work) the mean number of words in TOSAs are much longer. In fact, the PayPal TOSA Agreement revised as of July 9, 2019, runs about 90 pages. Although a substantial portion consists of charts listing fees that differ by country, there is still a great deal to review.

Reason #4: We've just thrown up our hands. In 2018, 53% of 1,000 people surveyed said brands knew too much about them, another 45% would part with their data for an average of just $151 a year to get a more personalized experience, 35% would allow Google to track and monitor them online for $23, and 9% would give up data for free. Some say we've traded privacy for convenience and efficiency, but Joseph Turow, a professor at the Annenberg School for Communication, says his studies show the reason is resignation: We feel that the privacy ship has already sailed, which is probably true.[101]

It does seem likely, though, that the laws regarding data sharing may be tightened. Allowing access not only to data from users but also to users' friends is part of the criminal investigation against Facebook begun in 2019. Access to friends has certainly been an exponential data windfall for those who took advantage of it. According to Statista, in 2016, 40.41% of users had 200 friends or fewer—but 38.35% had 200 to 500 friends, and 20.79% had in excess of 500. That's an enormous amount of data that could be mined to develop personality profiles and highly targeted messages.

Reason #5: We don't seriously consider the potential for harm. This is probably the biggest mistake of all.

The Dollar Costs of Giving Up Privacy

While everyone understands the direct fiscal dangers that the internet poses—having your identity stolen or your bank accounts hacked—many are less alert to the indirect but also significant financial consequences that may come from a loss of privacy.

And as technology becomes ever more sophisticated, your finances may be affected in additional indirect yet impactful ways. One disturbing possibility involves college admissions. To determine "demonstrated interest," some colleges are now tracking such things as how quickly prospective applicants are opening their emails and whether the applicants click on links to the school's events. A financial commentator suggested that if the applicant shows high interest, a school might be inclined to offer a very motivated student less financial aid on the supposition that person would stretch his or her budget to attend.[102]

The meme that "in the future everyone will be world-famous for 15 minutes" has come true, thanks to technology. Each of us can produce our own broadcast channel that can deliver sudden, productive, and even exhilarating results—many have made fortunes thanks to a YouTube video that went viral—but can also be your undoing. Not only can the social consequences be devastating. You may even become unemployable.

With one ill-considered, derogatory tweet, Rosanne Barr killed her career and inflicted collateral damage to her coworkers when her show was canceled (though it ultimately continued without her). Papa John's founder John Schnatter had to step down after the uproar created by his remarks about national anthem protests by football players (for which the company later apologized); at one point his personal stock holdings had dropped $84 million in value. A teenaged influencer lost millions of dollars' worth of sponsorships when her mother's involvement in the "back door" pay-your-way-in college admissions scandal was revealed, and the girl's imprudent remarks about her commitment (or rather lack of it) to a college education came under new scrutiny. And a formerly obscure Department of Transportation worker was suspended after she allegedly made remarks on Facebook calling for illegal immigrants to be shot at the Mexican border.

But you can become notorious without being a celebrity and jeopardize your financial prospects even if you're not a public employee. And it can happen literally overnight, as the director of corporate public relations at a tech company learned in a well-publicized incident.

Before embarking on a long international trip, the woman began tweeting snarky—intended as jokey—comments to her 170 followers. One tweet insulted a fellow passenger, a second took a poke at a country where she stopped en route, and a third had racist implications. Then, while she snoozed during the final leg of her journey, a blogger with 15,000 followers saw and retweeted the racist remark, noting the irony of a communications director's being so indiscreet.

Eleven hours later, by the time she landed, her comments had become the #1 worldwide trend, her employer had denounced her (and soon fired her), and tens of thousands had vilified her. She subsequently had a hard time reestablishing herself professionally. (Odd footnote: The blogger suffered his own public shaming a couple of years later when he posted a similarly tone-deaf "joke.")

What you reveal online may affect whether you get a career opportunity or promotion, a car loan or a mortgage. It doesn't have to be something explosive to change your future. It may be nothing more than the fact that you've been identified with a polarizing political position or an indiscreet remark. It may be misattributed or misinterpreted. But whatever it is, once something is out of the bag, it's hard to put it back in.

Never Forget That You're Posting for Posterity

What happens in Vegas stays in Vegas. What happens on social media stays on social media.

"I used to think this stuff was gone forever, junked to make room for shiny new facts," wrote blogger Daniel Koliz.[103] *Then he signed onto the Timehop* app. When you grant it access to your photos and various social media accounts, it culls a record of past posts. He was "cringing from a distance" when he spotted a decade-old photo informing the world he was "KING OF THE V-NECKS."

Those insensitive tweets, dumb photos, tasteless jokes, embarrassing emails, and TMI-revealing Instagrams have long-term potential to cause much more than embarrassment. A future employer may find one and see it as irrefutable testimony to your indiscretion or poor judgment, as we have seen in many public falls from grace.

The consequences may be even more dire.

If at work you make a comment that could be interpreted as hostile or discriminatory and if it is overheard and reported, an investigation may be instigated. Your entire digital past—not just emails but Facebook and all social postings—might be reviewed to search for instances of your sending, forwarding, or even responding "LOL" to what might be considered a tasteless or inappropriate comment or joke.

Potentially worse than being fired, you might be sued. We've all seen the movies and TV shows where the accused is read his Miranda rights by the arresting officers: "Anything you say can and will be used against you." You aren't protected from yourself once you've posted an online road map to who you are, accessible to manipulation and interpretation. You can bet on the fact that anything a hostile person can use against you will be used, and the electronic evidence might lead a person (or a jury) to infer that you have prejudices and rule against you.

Even benign postings may come back to bite you. You may put up tender affirmations of the exceptional qualities of your current significant other when similar comments about your previous relationships still exist online. A recruiter may someday come across both and think, *This person claims to know exactly what he wants, but see how fickle his behavior was in the past.*

You can't distance yourself from the narrative you yourself put online. If you fail to take into account that anything you post may come back to haunt you, you're not simply a victim; you're a collaborator.

NOTE: The world of technology is constantly evolving. The following information is accurate as we go to press, but some of the sites may have changed names, become irrelevant or obsolete, disappeared, or in some other way become unreliable. It's my hope that you'll get some good ideas and use what follows as the jumping-off point to do your own research.

Strategies to Protect Your Privacy

Check Terms of Service Agreements

You may not be able to plough through them in their entirety, but with a quick check you might be able to avoid some trip-ups. For example, a customer planning to switch to a bigger cell phone keyboard noticed that the terms said that in order to learn "your preferences," the manufacturer would have had permanent access to whatever the customer typed going forward as well as in the past. At minimum check:

- *Duration*: Does the company keep your data even after death?
- *Access*: Who will share data?

- *Encryption*: Is it in place?
- *Security breach*: What's the response?
- *Cancellation Terms*: If you cancel, is data deleted?
- *Rights and liabilities*: Who owns your data? Do you have a right to sue?
- *Change in terms of service*: Can they change the agreement? Retroactively? Without permission? Without informing you?

For more information about agreements and terms of service contracts, consult the following:

> Terms of Service: Didn't Read (tosdr.org) is a user rights initiative to rate and label website terms & privacy policies.
>
> Actiane Security Labs has created the EULA (End User License Agreement) Analyzer to assist people in better understanding the contracts they enter into while using software on the internet: https://www.spywareguide.com/analyze/.

Protect Your Passwords

- Don't use the same password in multiple places. If there's a security breach in a big company, you're vulnerable to being hacked somewhere else.
- It should be unique, long, and contain numbers (but not birthdates or other obvious clues).
- Use two-factor authentication (2FA). Most major web services support 2FA and can help to protect your accounts by adding another layer of security. In addition to your password, there are two other authenticating measures:

a device (such as cell phone) and something you "are" (a fingerprint). They should be using at least two. This would mean performing another action, like entering a six-digit code from an app after entering a password.

- Always change default passwords for online accounts and internet-connected devices such as routers and modems.
- Abine Blur is a suite of privacy tools that includes a password manager and masked email addresses. Just enter a real email address on the Blur website, and install its browser extension. Whenever you're asked to enter an email address, Blur pops up with a masked alternative. Emails sent to your masked address will be forwarded by Blur to your real address. Best of all, you can generate and destroy new masked addresses on the fly.
- Seriously consider using a password manager. It may take some time to set up, but you'll have peace of mind.

Get a VPN (Virtual Private Network)

- Without a VPN, your ISP (internet service provider) is fully open to anyone determined to steal your passwords, financial, and personal information.
- For a small fee, the VPN encrypts your data so it goes from your ISP to the VPN and then to the destination (be it video-sharing website, bank, search engine or other); the data is recorded as coming from the VPN rather than your device.
- VPNs are available for cellphones, but they don't work as well; however, phone data is harder for hackers to get into.

- A VPN is especially important if you're using public Wi-Fi, because you can't be certain if it's being operated by legitimate sources.
- If you need reviews, costs, and other information about VPNs, consult *PC Magazine*, reviews.com or another trustworthy source.

Protect Your Communications with Other Means

- The text messaging app Signal limits the data it collects and encrypts your messages. It's easy to use and free.
- You can also find encrypted backup services and encrypted file-sharing sites.

Check If You've Been Compromised

- Abine's Delete Me ($129 per year for one person, $228 for a second) looks within data broker and public records aggregator websites to find personal information for sale. (Of course, you have to provide a lot of personal information first.)
- HaveIBeenPwned.com aggregates the information from mass data breaches into a searchable service (including data that was disclosed by the companies and also public dumps of data from bad guys.) Type in your email to see which of your accounts were exposed.

Turn Off Location Tracking on Your Phone

- Check online to find out how to do it for your particular device.
- Facebook and Instagram still show location-based ads to users who have location services turned off. And

it isn't something you can opt out of. Facebook says, "When Location History is turned off (using Location Settings), Facebook will stop adding new information." But it adds, "You may still share your precise location when you use our products. For example, we may receive and store location data when you check in, RSVP as attending an event, or post photos that include location information."

Manage Your Cookies

- Apple, Google, and Microsoft have published instructions on how to clear data from their respective browsers: Safari, Chrome and Edge.
- Disable third-party cookies.

Block Whomever You Can

- Reset your advertising ID. While you cannot completely turn off the ad ID, both Android and iOS do offer ways to reset it. Find a means of resetting your advertising identifier as well as opting out of *some* personalized ad profile tracking.
- Stop Yahoo's Email Ad Mining. You may start by going to the Ad Interest Page.
- Delete ads from your history on Google. Go online to myactivity.google.com to find Google's stored data about you including the history of ads you have loaded.
- Use a private blocker on mobile devices. Privacy-centric mobile browsers like Firefox Focus, StartPage, and Ghostery Privacy Browser include built-in ad and tracker blocking. DuckDuckGo, probably the best known, doesn't use cookies and doesn't profile users;

it shows the same results to everyone. StartPage pays Google to source its searches but removes all tracking. These are handy when you want to do a discreet web search. (They may be impractical to use as everyday browsers because the built-in blockers can break important parts of websites.)

- Don't access public Wi-Fi unless you have a VPN.

Install a Tracker Blocker

Tracker blockers like disconnect.me detect snooping code on websites and prevent them from loading.

Delete Your Cell and SMS History from Facebook

If you have installed the Messenger App, then you will have given Facebook access to all your phone contacts. Facebook will then have access to all phone calls you make and text messages you send. Online, you can find instructions for deleting this information.

Check Your Dark Web Profile

Check to see if your SSN, phone number or email are on the dark web. Experian offers a free dark web scan: https://www.experian.com/consumer-products/free-dark-web-email-scan.html.

Consider Your Apps

Do they really need access to your microphone, contacts, or location data?

Shop Small and Local

Non-chain stores have less incentive—and capacity—to parse and sell your records.

Keep It Zipped

The less information you post online, the more secure you are.

It is relatively easy to recognize when we are being targeted with a marketing message from people who have determined our preferences for goods and services in order to attract our business and get our money.

More insidiously, people can in a similar fashion learn our biases and opinions from the data they collect. Microtargeting, as it's called, is beloved by brands and also by other kinds of influencers. Data combined with advanced machine-learning algorithms can build predictive models regarding a person's political views, religious beliefs, even drug use. When they use that information to manipulate our minds, the consequences are much more significant.

6

Surrendering Control

Is Technology Expanding Your View or Distorting Your Vision

We have begun to outsource our intellectual work to companies that suggest what we should learn, the topics we should consider, and the items we ought to buy. . . .The problem is that when we outsource thinking to machines, we are really outsourcing thinking to the organizations that run the machines.

Franklin Foer, journalist, former editor, *The New Republic*

THE EMINENT SCIENCE FICTION WRITER ARTHUR C. CLARKE wrote that he "would have believed anyone who told [him] back in 1962 that there would one day exist a book-sized object capable of holding the content of an entire library, [but] would never have accepted that the same device could find a page or word in a second." Such achievements inspired him to author what became known as Clarke's Third Law: "Any sufficiently advanced technology is indistinguishable from magic."[104]

Clarke died just a year after the 2007 introduction of the iPhone, an object considerably smaller than a book that accessed the contents of many libraries, and soon afterward, an uncountable number of them. Anyone with the use of a computer, smartphone, or other electronic device can tap into an amount of information that is already too huge to contemplate and that just keeps getting bigger.

By April 2019, internetlivestats.com estimated that out of about 1.675 billion websites online, 200 million were active. The sheer scale seems to have overwhelmed rather than enlightened us. Some say the result has been what is termed "infoxication": lots of data without insight, and perhaps worse, without purpose.

Clarke's Third Law remains prescient. We continue to impute to technology an omniscience and power that make us too reliant on it, too willing to accept without question the instant answers it produces, and too quick to assume that merely having access to it gives us special capabilities. And instead of our having agency over the tool, it seems that we are giving it agency over us.

A few years ago, the Pew Organization asked 1,000 people about the utility of the Internet and cell phones. Rather than feel overloaded by information, 72% enjoyed having the access. What's more, an overwhelming majority—87%—said both these tools had, over the previous five years, improved their ability to learn new things.[105]

What "new things" were they learning? More than 70% mentioned international news, national news, and pop culture, and more than 60% got information about hobbies, personal interests, and their friends. But the top category people mentioned was products and services to buy. It seems that we're predominantly using these magical tools to get shopping tips along with celebrity news, sports coverage, and updates on our pals. The internet offers us an unlimited menu of choices, and we select the equivalent of junk food.

Having information and options doesn't translate to having insight or judgment. Just look at the subprime meltdown. Taking advantage of cutting-edge technology, some of the largest banks and financial institutions fed it some faulty assumptions and it produced elaborate financial products that turned out to be toxic. It was the ultimate example of garbage in, garbage out. What they did was synthetically engineer the foundation for the Great Recession.

Are You Seeking Information or Is It Seeking You?

In a complicated world, we need expert assistance from the people who provide us with the necessities to keep us alive and safe as well as judicious counsel from the professionals on whom we rely to protect our interests, our health, and our fiscal wellbeing. We expect them to stay informed and current, distill what is useful and practical, and apply what is relevant to us—their customers, patients and clients.

But no one benefits more from being alert to what is useful and informed about what is necessary than we ourselves. What the internet has given us is an unimaginably powerful tool to make the job easier than ever, yet we aren't especially diligent about using it.

Not so long ago, doing research meant going to libraries, thumbing through card catalogues to find relevant books and then searching for them on the shelves. To find newspaper and magazine articles, you'd have to pore over each of the annual volumes of *The Readers' Guide to Periodical Literature*, though few libraries had anything but the most popular magazines, and recent copies only. If you were lucky enough to find microfilmed copies of more obscure or older issues, you'd load a spool of the film on a viewer and crank your way through the reel to search for a specific article in a specific issue.

Today, we just go to Google. We are so trusting of it (and other search engines, though to a lesser degree)—and, let's face it, so lazy—that more than 90% of people don't go past the first page of the search results. In fact, more than 50% don't go past the first two results.

That's not wise. Studies have proven that any item that appears at the top of any list commands outsized influence. Many Search Engine Optimization (SEO) companies promise their clients that they can "guarantee" a top ranking. While Google offers an online video that gives advice about choosing an SEO company and legitimate ways to boost rankings, it insists that a "guaranteed" top ranking is impossible. Yet what appears at the top of any list is not always the most relevant or unbiased result, since many factors are involved.

In searching for a restaurant recommendation in an unfamiliar neighborhood, for example, you may be unaware that on the Google food finding app, a restaurant's relevance, distance and prominence may influence its ranking more than the quality of the food. You're more likely to be referred to a fast-food franchise than an Ethiopian café that's a local gem. And some places may get your attention simply because the owner has paid the site owner to include photos or to ensure that all competitive ads will be nudged from the page.

By looking only at what we are led to, we don't spend a lot of time considering what we might *not* be seeing. This can have major consequences.

In 2015 the Proceedings of the National Academy of Sciences published a paper called "Manipulation," which reported that biased search information could shift undecided voters' preferences by 20%. "Because the majority of people in most democracies use a search engine provided by just one company," the authors wrote, "if that company chose to manipulate rankings to favor particular candidates or parties, opponents would have no way to counteract those manipulations."[106] At the time, this might have seemed unlikely, but today it certainly does not.

Just about everything posted on the web is there to serve someone's purpose. Never forget that.

Look Behind the Curtain

When you land on a site, it's often obvious that its intention is to sell you a product or a service. But the motivation behind a site that seems to be offering information is not always clear. Sometimes it is genuinely a public service, but often it's designed by a person, an institution, or an organization with a specific message meant to influence your thinking.

True, that's the case in other forms of media, but elsewhere, the intention is usually made clear. Advertising is usually readily identifiable as such on radio, television and in print with the announcement "brought to you by," the word "advertisement" on the page or (although less obviously), as I previously mentioned, accompanied by the hashtag #ad. Most viewers are generally aware of the intent behind widely-known platforms—the opposing perspectives, for example, of Fox News and MSNBC and the different points of view held by *The Wall Street Journal* and the *Washington Post.*

Things are murkier on the web. With so much material available, you can't always determine, especially when you're making a quick assessment, which site is the product of an expert source and which is coming from an uncredentialed content creator. It's not always immediately clear what position the source takes on issues you care about and whether you're getting a complete or balanced picture.

Remember when Dorothy in *The Wizard of Oz* pulls back the curtain to discover that the "wizard" is not the omnipotent force she had imagined? It's important for you to take a peek behind the screen to determine where commentary or information is coming from. Is the source a disinterested (meaning *unprejudiced*) individual or a disgruntled customer making a revenge post, a relative endorsing a cousin's business, a public relations firm whitewashing a client's misstep, a foreign operative trying to influence popular opinion, or some other individual or entity with an agenda to put forward or an ax to grind?

Even facts can be interpreted differently according to the bias of the presenter. One group of researchers writing up an experiment about dealing with back problems may conclude that exercise is the solution, and another will suggest that certain vitamins or minerals are the answer. Searching for additional information, you may find that the first group is on the payroll of a gym equipment company and the other is funded by a company that manufactures supplements.

It would be unsurprising if a financial expert who works for an insurance company would present solutions to clients that all involve the use of an insurance product or that one who works for a commodities company might recommend gold as an investment choice. As the saying goes, to a hammer, everything looks like a nail.

Even what seem to be well-intentioned or neutral sources may not be. People were generally shocked to learn that Facebook's Trending News feature introduced in 2014, reported to be an attempt to woo Twitter users by affording a quick look at the news and purporting to present the most popular topics being discussed on the platform, was actually "curated" by editors with a progressive bias selecting the stories they considered newsworthy.

Journalist Franklin Foer pointed out the very critical difference between journalism and curated internet "news." Unlike journalism, which is founded on the principle of giving people what is good for them, Facebook's goal was to give people what they *wanted.*

While social media generally used "time spent" on its site to measure the quality of engagement, for its Trending News feature, Facebook used an algorithm that gave higher value to "meaningful" interactions and engagement. Its criteria were based on whether a post got comments and reactions ("likes," "Haha," etc.) and, in addition, whether there were replies to the comments, whether the post was shared, and whether the shares got comments.[107] And if they wanted more video, fewer words, or material that pleased them ideologically rather than hard news, they'd get that too.

Trending News was shut down in 2018. But today content everywhere is often determined by similar means.

Getting Information vs. Getting Validation

All online content aims to be predictive. Every platform knows that the best way to attract you is with something that aligns with your past preferences, whether it's an image, a hard news story, a music video, or celebrity reporting—and content creators use algorithms' "rules" to make such determinations. Your media becomes a product of what you already like, as filtered through a tech company's assumptions about what you already like, which they gather by collecting data from you.

The results can be highly personalized. For example, Netflix learned that artwork commands 82% of viewers' focus, even within a favorite category, so it's the factor most likely to influence what a viewer chooses to watch. Since its consumer base is too large for any single image to have universal appeal, Netflix introduced a "recommendation algorithm" to show viewers images based on their previous choices.

So if your recommendations from Netflix included "Good Will Hunting," and in the past you showed a preference for romantic comedies, you'd see a picture of Matt Damon and Minnie Driver kissing. But if you preferred comedies, you'd be shown an image of Robin Williams. The Netflix algorithm keeps gathering information about each viewer, changing the images as it gets to know the viewer better and better.[108]

The two basic concepts underlying machine learning are exploitation and exploration. *Exploitation* gives viewers selections based on their past preferences. *Exploration* introduces them to new ones.

Netflix encourages exploration for its own very pragmatic purposes. It costs Netflix more to offer major hits, so it has a financial motive to lead you to more obscure titles to increase its profit margin. Also, it wants to motivate you to sample all the wares on its menu.

But when algorithms bring you your news, content providers are more likely to lean toward exploitation, bringing you more of what you have preferred in the past. That keeps you returning to the platform. It also reinforces your beliefs and effectively shuts out other points of view.

Opening Your Eyes or Closing Your Mind?

The invention of the internet excited giddy expectations. Optimists hoped that the huge, global forum would permit people to exchange and access ideas and opinions on a tremendous scale and make them more open-minded. Instead, the opposite seems to be true.

Before the internet, media kept us informed through journalists' reports and editors' decisions about which stories to print or broadcast and how much prominence to give to each. Technology makes it possible to get information produced for us based on digital or algorithmic data mined from our own online travels. Motivated parties "know" a great deal about each of us and can use it to affect not simply what movie or lamp we choose but decisions of far more consequence.

If on one computer you consistently search stories with the theme that "Republicans are great," and on another you search the contrary theme, "Democrats are great," within a month the news stories you get will likely be skewed to one viewpoint or the other. That's because over time, you and the technology become one. While you may hope to get a balanced picture, information generated by algorithms has filtered through massive amounts of material to decide what you *want* to see—a manufactured permutation of your own opinions. The websites you encounter will reinforce your political beliefs and exclude what isn't in alignment—becoming what have been termed "online echo chambers powered by algorithms."

Reviewing Tom Nichols' book *The Death of Expertise*, critic Michiko Kakutani wrote that the internet has given people "the illusion of knowledge when in fact they are drowning in data and cherry-picking what they choose to read. Given an inexhaustible buffet of facts, rumors, lies, serious analysis, crackpot speculation, and outright propaganda to browse online, it becomes easy for one to succumb to "confirmation bias."[109] This is a well-recognized phenomenon: the human tendency to pay attention to—in fact, seek out—information that confirms and strengthens our beliefs and to ignore or minimize the importance of information that contradicts them.

A study about climate change, a hotly debated issue, illustrated confirmation bias in action. To a randomly composed group who had a mix of attitudes about climate change—some strong believers that it was man-made, some deniers, and some in between—researchers delivered news that the situation was more dire than originally thought. Subsequently, the researchers told the group that another report showed that the opposite was true. To deniers of man-made climate change, the comforting news of the second report had a big impact, but the initial alarming news did not. Strong believers reacted in precisely the opposite manner.

We are constantly exposed to competing claims about many other issues; among them are affordable care, immigration, terrorism, and increases in the minimum wage. As the sheer volume of information increases, polarization will be heightened. And as we get news that's more and more curated, meaning more *infotainment* than *information,* it's harder for us to sort out what's real and what's not.

To illustrate the consequences of confirmation bias, blogger James Clear drew a Venn diagram of two overlapping circles, one labeled "What the facts say" and the other, "What confirms your beliefs." He labeled the first, "Undervalued," the second "Foolish," and the area where they overlapped "Overvalued." The more you think some source validates you, the easier it is for that entity to take advantage of you and manipulate you.

Technology and the Rise of Fake News

While there is a recent tendency to use the term "fake news" to describe anything that doesn't align with your own perspectives and beliefs, I am talking here about news stories that appear to be an oxymoron: "genuine fakes." In other words, total fabrications.

Cameras and recording devices are now so advanced, affordable, and ubiquitous that documentation of events is more common and easier than ever. And "seeing is believing." Or is it? Technology has also brought us tools that make it very simple to alter recordings and pictures. Matt Groh, a researcher at the M.I.T. Media Lab, built a system called Deep Angels that can remove people and objects from photos. Even videos can be doctored so that the alteration is undetectable, making it possible to "prove" any narrative. Technology futurist Aviv Ovadya, Chief Technologist at University of Michigan's Center for Social Media Responsibility (UMSI) warns of a potential "infocalypse," a technology-enabled misinformation crisis, when "anyone can make it appear as if anything has happened," such as a foreign leader making a declaration of war, even when this did not take place.[110]

In fact, you don't even have to doctor the visuals. You can simply allege that the ones that contradict your position have themselves been doctored. That's because "truthiness" is increasingly equated with truth, and fact-checking has little value to people who want only confirmation of their own beliefs. It seems that lies have become just another useful tactic disseminated by political operatives who weaponize them, public figures with a secret to hide, and companies attempting cover-ups of their missteps.

After all, it is not so difficult to succeed in spreading information contrary to fact. For one thing, researchers have found that two people looking at the very same image (video, document, photo) can come away with two very different notions of what they've seen, which explains why public figures can successfully "walk back" comments that hundreds of millions of people have watched them utter in real time.

And technology effectively reinforces our biases. The internet is a vast copying machine that users employ to rapidly distribute memes—an image, video, or piece of text—that may be humorous, touching, or exciting but can also be dark material like a conspiracy theory, a false rumor, or dangerous lie. As might be expected, the more surprising or bizarre the meme, the more likely it is to be passed along. Sometimes people send a meme in earnest (because it validates their opinions), but they may also be doing it ironically (because they think it's obviously ridiculous).

Unfortunately, once digital social networks register that people find a particular meme appealing (the algorithms of course make no distinction as to whether it's because they find it compelling or consider it absurd), they give it more traction. More and more people start sharing it, although many (perhaps most) give it only a cursory read. Once memes start to spread, it's hard to correct an incorrect message because the memes move too quickly for fact-checkers to keep up.

To support its claim of neutrality, in 2015 Facebook released a study by its own scientists that was published in the journal *Science.* It found that "individuals' choices played a stronger role in limiting exposure" to contrary news and commentary than Facebook's own algorithmic ranking.[111] But others disputed that. One found the opposite to be true: that homogeneous online networks help conspiracy theories persist and grow online, creating "an ecosystem in which the truth value of the information doesn't matter . . . All that matters is whether the information fits in your narrative."[112]

One study of the 2016 election found that only about 8.5% of Facebook users shared even a single fake news link (defined by the study as "news disseminated by a limited number of sites that reliably, according to various sources, published fake content"). Prolific Facebook users were less likely to post fake stories, lending credence to the theory that more active and experienced users could discern which news was fake. Yet it seems likely that in the absence of checks, fake news will continue to be more pervasive and ubiquitous.[113]

For one thing, certain cognitive processes help facilitate the spread of fake news. Dr. Colleen Seifer, a professor of psychology at the University of Michigan, explained that when you hear a news report, first your mind must grasp the main concept. For example: Clinton=sex offender, Trump=racist. You may subsequently refute that first impression, but it may linger and can be reinforced by repetition—even if it's initially presented as a joke.[114]

The human mind is extremely susceptible to forming memories, which are hard to erase. And since the process of remembering one thing can cause us to forget another, fake news, by seeding the population with false memories, can potentially alter our recollections of the past.

Also, when information comes from friends, it's overweighted, which is especially significant given that nearly two-thirds of Americans get at least some news from social media. Therefore, it's alarming that the tech futurist Aviv Ovadya warns of the future use of AI to scan things like your social media presences and craft false but believable messages from people you know. Once you have been influenced by the data delivered by algorithm and by your friends' opinions, it takes a lot of motivation to check the facts even if a story may seem questionable.

In 2018 Steven Brill, author and the founder of the magazine *The American Lawyer*, and Gordon Crovitz, a former publisher of the *Wall Street Journal,* introduced NewsGuard (newsguardtech.com), "whose trained journalists will assess news websites for credibility and transparency," assigning a rating for trustworthiness (green) or not (red) and giving "nutrition ratings" for 2000 news and information sites. Once you install NewsGuard's free browser log-in, NewsGuard rating icons will begin to appear next to links on search engines and social media feeds, including Facebook, Twitter, Google and Bing.

"The most important thing we've done is demonstrate to 90% of the public that there actually is a way to do what we're doing with human intelligence instead of algorithms and do it apolitically," Brill said. He admits that a site can get a 100% rating and still unwittingly publish false stories from time to time, but NewsGuard is making the effort to provide a useful consumer's guide. It's probably good news that a second site with a similar intent, Credder, debuted in mid-2019.

How Social Media Moves Markets

How does all news—fake or real—affect your nest egg? As I noted, information from a friend is given extra weight. But our definition of "friends" has been modified by social media.

Facebook, for example, has a two-way system. Both parties have to sign on to become friends. But on Twitter, the system is more complex: You can be a follower of someone; someone can follow you; you can both follow one another; or neither of you can follow one another.[115]

A lot of people who are "followers" of people who are well-known may conflate that relationship with being "friends" of those people, so they may be more likely to pay attention to their tweets. That is even more likely when the well-known person's large number of followers validates that person's importance. I mentioned previously that many influencers may in fact be bots. It also seems that many "followers" are about as real as the imaginary playmates of your childhood.

In November 2017, Facebook revealed it had perhaps 60 million fake accounts, or "bots." In January 2018, *The New York Times* reported that a firm called Devumi, along with some smaller competitors, had made a business of selling Twitter followers and retweets—some 200 million followers-for-hire in all, pulled from three to five million estimated accounts, some belonging to active Twitter members (fraudulently repackaged by Devumi) and some not.

Twitter suspended more than 70 million fake accounts in May and June 2018, and deletions continued at the rate of one million a day. In July, Twitter said it would stop counting "locked" accounts as followers, in an effort to make its user data more accurate. Twitter lost 7.8 million followers in July but regained 2.36 million of them by mid-October, when many accounts appeared to have been unlocked—which can happen after a password reset—according to data from the ad fraud consultancy firm Social Puncher. The firm told Reuters that it suspects the affected locked accounts are controlled by fraudsters who sell followers to artificially boost an account's popularity.

In November of 2018 it was reported that Twitter again deleted millions of followers who had reappeared after the summer purge. As of this writing, it's unclear if Twitter can reliably distinguish between fake and legitimate followers.

The customers for fake followers are primarily people who could turn big followings into money, influence, or both. These include celebrities who could command more bookings and higher salaries, would-be authors who might secure publishing deals, marketers who wish to appear more reputable and important, and the influencers I've discussed who attract sponsors and command higher fees for touting certain products. For them, it's about personal gain.

But a person or institution with another kind of agenda may use social media to disseminate fake news with impunity. In one instance, a video of Speaker of the House Nancy Pelosi, doctored and slowed to make it appear her speech was slurred and thus raise questions about her health, was released on Facebook and shared millions of times even after Facebook was alerted that it had been altered. A Facebook representative responded that the social network's policies didn't stipulate that postings had to be true. A post could be removed only if it violated the social network's community standards, which at that point (in May 2019) were silent regarding how to handle videos that were manipulated in a way that created a false impression.[116] (The post was ultimately removed, though no source took credit for that.)

We saw the immediate and far greater consequences that fake news can produce in 2013 when someone hacked into the AP Twitter feed and reported that President Obama had been injured in an explosion at the White House.

This was a classic demonstration of how a news story becomes a market mover in the blink of an eye. For high-frequency trading, computers collect data from a variety of news sources, including print and broadcast media, SEC filings, business publications, and even (as this incident demonstrates) Twitter. They look for specific words or phrases that might have a business implication, such as "takeover," "merger," "acquisition," and so on, and the names of companies and words such as "Barak Obama," "White House," and "explosion." Computers digitize this information as soon as it is available and feed it into other computers, where it initiates algorithm-driven activity that often results in enormous market swings.

Coming as it did a week after the shocking Boston Marathon bombing, which killed and maimed hundreds, the report of the then-president's injury was accepted as plausible by a jittery public, and it seemed even more authentic since it had come from a respected source. Consequently, it was retweeted instantly and widely—about 4,000 times, which just added to the story's credibility.

Although the AP worked fast to get out the word that the news item wasn't true, its impact was huge. In seconds, the fake news caused a .09% decline in the S&P 500—$130 billion in market value.

Fake financial news is usually created for a very specific purpose: to drive traffic to a site, to get people to sign up for a program to solve the "problem" the fake news is reporting, or to scam people to get at their money and personal data. This event fell into the last category as a deliberate attempt to gin up the market and create turmoil on behalf of traders who make money on short-term momentum.

There used to be a myth that if you were shorting a stock (betting that it would go down), you should go into one or more of New York's biggest brokerage firms, get into an elevator, and drop an envelope on the floor marked "PERSONAL AND CONFIDENTIAL" that contained a phony document with a very negative report about the company you were shorting. A sell-off would soon follow. The internet allows you to do a monumental, accelerated version of the same scam. A fake story that moves the market for less than a minute can reap millions for people who are able to make trades in nanoseconds.

In another such case, a January 2018 Bloomberg News article reported that China was slowing or halting its purchases of U.S. Treasury bonds. Within an hour, yields on 10-year U.S. Treasury bonds jumped from 2.54% to almost 2.60%, double the average move. Later in the day, the AP reported that China denied the earlier story.

On reflection, the news made no sense. China was holding over $1 trillion in US bonds; dumping them would have decreased their value. And buying the Treasury bonds makes the US dollar stronger and weakens the yuan, so Americans buy more exports from China. The dump would have been bad for China's economy.

The move might have been a Chinese ploy to show it can affect the US markets or an action by someone shorting Treasuries. Someone made money, but it wouldn't have been you.

Technology Heightens the Impact of News

Of course, real news can send the market into turmoil as well—particularly if it's perceived to have adverse economic significance or seems widely disruptive or disturbing. While negative news has always had a negative impact on the stock market, technology has increased its effect, both in terms of speed and magnitude.

Among the many unsettling headlines we have seen in recent years, those that most notably rattled the markets were about falling oil prices, the fiscal crisis in the United States, Greece's threat to default, and the shocking Brexit vote in favor of England's leaving the European Union.

In all those instances, the algorithms that dominate the market's daily trading volume started pushing all stocks sharply lower, without any meaningful selection of the companies being sold—a perfect example of uncontrolled short-term volatility.

The day after the Brexit vote, there was another example. The Dow Jones Industrial Average dropped 500 points in early trading and the S&P 500 lost 1.7%. Less than three weeks later, the

S&P 500 had risen beyond its pre-Brexit levels and reached an all-time high.

The US-China trade war created volatility for months in 2019.

Following are some of the headline stories and the movement they caused in just one three-month period:

> 7/01: US says no more additional tariffs on China.
> DJIA up 117, S&P 500 up 23
>
> 8/01: President Trump tweets of added 10% tariff.
> DJIA down 280, S&P 500 down 27
>
> 8/13: White House announces changes in scope, dates of tariff.
> DJIA up 373, S&P 500 up 43.
>
> 10/08: US blacklists 28 Chinese corporations, increasing tariffs.
> DJIA down 313, S&P 500 down 45.
>
> 10/11: Positive reports about trade negotiations.
> DJIA up 320, S&P 500 up 32.

This pattern has become somewhat familiar. A sharp or subtle move up or down is followed by concerns over whether the trend will continue or reverse. Then the pundits weigh in, creating an agitating thrum of mixed messaging that provides no comfort: "This is important." "It may continue. Or it may not." "It may be good. Or it may be bad." "But pay attention."

24/7 News Is Bad News for Investors

By many measures we are at the moment living in one of the safest, happiest, healthiest times in human history. There are exceptions, of course, but this is true not just here in the United States but also around the world. As of 2019, there are 90 democracies—up from 11 in 1900.[117] Global literacy has increased from

42% in 1960 to 86.3% in 2019[118] during a period when the world's population went from 3 to 7.7 billion.[119] In 1990 more than a third of people in the world lived in extreme poverty—living on $1.90 a day or less. In 2015, the most recent year with robust data, extreme poverty fell to 10 percent, the lowest level in recorded history.[120] The Brennan Center for Justice reported in 2019 that crime in the United States had hit historic lows, and there are similar reports from almost all over the planet. Life expectancy at birth for the world's population was 72.6 years in 2019, an improvement of more than 8 years just since 1990.[121]

By 2014, the life expectancy in this country had gone all the way up to 78.9.[122] In this respect, as in many others, the vast majority of people in the United States have lots more to be grateful for than most of the people in the world. Still, in 2017 life expectancy here declined for the second time in three years (after stagnating in 2016 and falling in 2015) because of rises in opioid use and suicide.[123]

Also, diagnoses of major depression rose by a third between 2013 and 2016, according to a 2018 report from Blue Cross Blue Shield Association.[124] And those findings were based only on data from 41 million health records of people who had actually been diagnosed with depression. The actual number is no doubt much higher, since many people with symptoms of depression have never been diagnosed or sought treatment.

There's other proof that Americans are worried and fearful. The American Psychological Association published a "Stress in America" survey in 2017. Some 63% of the 3,400 people they surveyed in 2016 considered the country's future a "significant source of stress," and 56% were stressed by "the current political climate." In the 2018 edition of the survey, the numbers jumped respectively to 69% and 62%.[125]

Humans are wired to experience fear, scientists say, in order to survive. Our ancestors learned a lesson when they saw one of their own attacked by a wild animal. Images and stories are powerful ways to teach fear. Thanks to technology, with our 24-hour-a-day, 7-day-a-week news feeds, we've got plenty of both. Although there are various explanations for the reason people are turning to drugs, considering suicide in ever-increasing numbers, and feeling depressed, it seems likely that a substantial part of the problem is our near-constant exposure to news.

What constitutes "news" is something unexpected and likely to be unwelcome. Media don't cover stories about planes that land safely or daily temperatures that are in the normal range. If the lead story is being broadcast from the beach, it's most likely because someone was attacked by a shark. And since the all-news channels are competing with entertainment platforms, they are even more likely to focus on attention-grabbing (and anxiety-producing) topics.

Events that might once have escaped our notice or that we might have learned about belatedly now get our attention as they are happening. We see the pictures and hear the details in our homes and offices, at the health club, in the supermarket, while waiting for the doctor, on the phones we carry in our pocketbooks and in our pockets, and even on the watches on our wrists.

Creating enough content to fill continuous news cycles is a challenge. To make their stories bigger and more compelling, print and broadcast reporters and editors don't simply deliver the news. They also comment and speculate on it, at length. In fact, almost immediately after any story breaks, the news shows round up a panel of experts to make an analysis. That can quickly lead to emotionalizing and sensationalizing. Playing up the potential negative outcome of a story makes it more compelling.

When a rare tornado struck in New York, the story churned up a lot of anxiety about human-caused climate change followed by apocalyptic predictions about the future.

Just the way in which a story is presented may affect your perception of its importance. A report that the Dow "plummeted" 250 points is likely to create more anxiety than a statement that "Today the Dow declined 1%." Longer stories appear to be more important than short ones, and something introduced as "breaking news" may seem to be especially significant; but the length of a story or how it's presented may just be the editor or news director's way of luring an audience.

It's very human to react instinctively to any alarming news, but this can do great damage to your nest egg. As an investor, you should always leave some time between hearing bad news and reacting to it. For one thing, by the time you start running for the exits, chances are you'll be way at the back of the line. For another, if you've been thoughtful in picking your holdings, wait out the crisis and are patient, chances are you'll come out ahead. Historically, the stock market's success at recouping losses and ultimately surpassing its previous high is 100%.

Part Three

TACTICS

7

Staying Relevant

Upgrade Yourself Continually

The illiterate of the 21st century will not be those who cannot read and write, but those who cannot learn, unlearn, and relearn.

Alvin Toffler, author of *Future Shock*

MY WORK IS ALL ABOUT THE FUTURE: collaborating with clients to build their assets and anticipate factors, known and unknown, that may affect their financial long-term well-being. In the former category are our health and career prospects. In these areas, as well as everywhere else, technology has brought about huge changes. The challenge to us is to see the opportunities they offer.

Medical advances have dramatically extended the number of years we are likely to live. But it's up to us to try to ensure the quality of that life by taking care of our health and managing our finances to preserve our lifestyle for our lifetime.

While many of us look forward to retirement, a great many hope to extend their years in the workforce as well, both for the economic benefits and for personal satisfaction. That also requires personal effort. Technology has changed the very nature of many jobs and will continue to reshape every aspect of work.

Futurist Yuval Noah Harari, author of *Sapiens: A Brief History of Humankind* and *21 Lessons for the 21st Century,* predicts that what lies ahead will be a cascade of ever-bigger disrupters. To be prepared for more change, we have to make some changes of our own, to unlearn some old things, relearn others, and learn some new ones.

We have already become accustomed to regular, incremental upgrades to our operating systems and software which go from OS 1 to 1.1, 1.2, and so on. That's a model for how we ourselves have to keep up: learning and adapting so that You.1 becomes You.1.2 and perhaps at some point You.2 and beyond. It's challenging but essential to move ahead so as not to get permanently left behind.

Will Robots Put Us Out of Work?

Artificial intelligence has already begun to change the nature and scope of jobs we thought were the sole province of humans by handling tasks in areas that a short while ago would have seemed unthinkable—like journalism. At Bloomberg News, a Cyborg system assists reporters in preparing articles on earnings reports. It "reads" a financial report and turns it immediately into a news story with the necessary facts and figures. Robot reporters have produced articles on minor league baseball, high school football and earthquakes. During the Olympics, a messaging system was set up to alert editors if a competition result was 10% above or below an Olympic record; such a difference might be significant enough to demand the attention of a human writer who could draft a longer story.

Document review, which for law firms has traditionally reaped many billable hours and kept paralegals busy, can now be done by software programs that scan legal papers to select those that are relevant.

Recently, a startup company introduced a program that can perform a magnetic-resonance imaging analysis of blood flow through a heart in 15 seconds, far more efficiently than human radiologists, who require 45 minutes to do the same.

Scientific American has reported that surgeons are already using robotic assistants to remove damaged organs and cancerous tissue, and a STAR (Smart Tissue Autonomous Robot), challenged with repairing the severed intestines of a live pig, outperformed human surgeons.[126]

Nevertheless, in a TED talk, economist David Autor says it's a mistake to be mourning the death of work. Half a century after the A.T.M. was introduced, there are more humans working as bank tellers than ever. But the nature of the job has changed. Rather than do rote work like counting out bills, they focus on more cognitively demanding tasks like developing relationships with customers, problem solving, and introducing them to new products.[127]

The goal is not to replace human employees. It's to free them to spend more time doing what they're good at: creative work, storytelling, digging for information, thinking critically, and exercising judgment. These humanistic skills have traditionally been honed as part of a college education, which for a long while has been regarded as a ticket to a solid middle-class (or higher) future. But escalating costs have raised a lot of questions.

Is a College Degree a Good Investment?

The cost of higher education has risen precipitously. Look at what happened between 1998 and 2019 to the cost of tuition alone:

- Tuition and fees at private universities on average jumped 168%, from $16,253 to $43,562.
- Out-of-state tuition and fees at public universities rose 200%, from $8,440 to $26,528.
- In-state tuition and fees at public universities increased a staggering 243%, from $3,168 to $10,858.

The website https://www.chronicle.com/interactives/tuition-and-fees shows costs for the present academic year (including tuition, room and board) and what the same package cost 20 years ago. The increases are jawdropping.

Here's a specific example: In 2000, friends of mine paid $30,450 for their freshman son's tuition plus room and board at a private liberal arts college in upstate New York. When their nephew was accepted to enter the college in 2019, the tab had risen to $68,768 (and that was for the most modest of three tiers of dorm accommodations). In addition to those basic costs, related necessities such as books, supplies, transportation, and other expenses jack up the required outlay.

Those numbers are increasingly out of reach for most people because they're so out of line with income growth. According to Pew Research, from 2000 to 2019 weekly wages rose just 4.3% (in real terms) among workers in the lowest quarter of the earnings distribution, and even among those in the top tenth of the distribution, wages rose just a cumulative 15.7%.[128]

The innovative thinker Scott Galloway, a business professor and founder of the brand strategy firm L2, has proposed that one or more of the "The Four" tech giants Amazon, Apple, Facebook, and Google absorb the costs of launching a tuition-free university that blends online and offline learning, pick off the top students for themselves, and charge other firms to recruit qualified employees from the remainder. Great idea.

In August 2018, Ken Langone, Home Depot founder and member of the Board of NYU Langone Medical Center, announced that he and his wife were making a $100 million gift that would enable NYU Medical School to offer free tuition for every entering student, and a Morehouse alumnus giving a speech to graduating seniors in 2019 offered to pay off the college loans of everyone in the class.

But both gifts will benefit just a very lucky few. Many people are staggering under the weight of college debt, now a $1.5 trillion crisis.

A lot of parents are preparing to shoot a bazooka through their savings in order to pay college costs. Okay, it's true that a college degree is a prerequisite to 35% of jobs openings today. And the National Center for Education Statistics says adults who complete a bachelor's degree on average earn 57% more than those who have only a high school diploma.

So a college education at least for the time being seems like a worthwhile, perhaps obligatory, investment in the future. But I suggest that parents look at it as they would any other investment: What's the payoff? Are you picking up the tab for a very expensive "brand name" college for bragging rights? Or do you think the school will generate higher future earnings?

As a matter of fact, research has found that—in terms of lifetime earnings—upper-middle-class students who have declined acceptance at top-tier schools and instead enrolled at less selective ones in the long run do quite as well as cohorts who graduate from more prestigious institutions. That's because their privileged backgrounds have given all those upper-middle-class kids the network and connections that help ensure success. Adding a top-tier school degree didn't make a difference. The network/connection factor also explains why children who come from middle-class families earn 162% more over their careers than children of families of lesser means.[129]

Young people who are assuming their own college debt should also think prospectively. I counsel them to focus on the total obligation they're taking on rather than think only about their monthly payments. The former is a lot more daunting. The federal student loan interest rate for undergraduates for the school year 2019-2020 was 4.53% and for unsubsidized graduate school loans was 6.08%. By comparison, mortgage rates were hovering around 4.0% for 30-year and 3.375% for 15-year loans. And mortgages are easier to refinance.

In addition, unlike mortgages and other loans, the college loan obligation is not forgiven even at death. (Your estate, or your cosigners and their estates, continue to be on the hook.) Promises of forgiveness if you spend ten years at a not-for-profit institution don't always seem to materialize, and it's always possible Congress will end this program (as nearly happened in 2019), and you'll still be stuck. If you hope to buy a home or start a family, the effect of a huge college debt on your credit rating may hamper or even trash those plans.

And here's the worst possibility: that you may not finish your degree, which is the case with 40% of students today. Dropping out is one of the most egregious mistakes anyone can make. Not only do college dropouts earn only slightly more per year than a person with a high school degree ($38,376 vs. $35,256), but also their being saddled with loan debt means they'll have less left to spend from each paycheck.

Bottom line: If you aren't absolutely determined to complete your degree, taking on debt for college is a risky bet. And even if you do complete the degree you may not have a significant advantage over those who don't. Since 2000, one study showed that 25% of those who do manage to complete the degree earn no more than the average high school graduate.[130] Perhaps it's because so many people earn a degree nowadays that simply having one no longer makes you special.

What *does* make a difference in earnings is the major you choose. Georgetown University was reportedly working on a four-year program that would combine a liberal arts bachelor's degree with a vocational master's degree, giving you both a traditional college education and practical job skills in four years. But until that's available, it pays to be strategic about choosing your field of study.

A Georgetown study says the lifetime earning difference between the highest-paying and lowest-paying college majors is $3.4 million.[131] Your major isn't necessarily your ticket to the big time. While in general, engineers far outearn teachers, the top 25% of education majors earn $59,000 or more annually, while the bottom 25% of engineering majors earn $59,000 or less. However, almost all of the highest-paying majors are in business and engineering, and the lowest are in education (plus arts and social work).[132]

An advanced or professional degree is also a good investment. A study a few years ago reported that lifetime earnings with a master's degree were $2.5 million, only slighter higher than with a bachelor's degree ($2.1 million), but those with a doctorate could expect to earn $3.4 million, and with an MD or JD, $4.4 million.

And again, what you study also makes a difference. There are rapidly expanding opportunities in the area of exponential technologies. This includes the whole spectrum of innovative breakthroughs, inventions, and discoveries that are progressing at a pace with or exceeding Moore's laws, including artificial intelligence, advanced robotics, virtual and augmented reality, alternative energy systems, biotechnology, and digital medicine.[133] This turn of events has created a market demand for skills in data science, supply chains and manufacturing while lowering demand in other fields.

Although 85% of US employers expected to hire MBA graduates in 2018, that was a 6% decrease from the previous year. Instead of business majors, both startups and established companies are increasingly hiring technologists, developers, and engineers. The really hot category is data analytics. In 2018, 53% of employers worldwide expected to hire graduates with a master's in the field, a 35% increase in demand.

These trends have not gone unnoticed by students. From 2006 to 2017 enrollment in *specialized* graduate business programs doubled. Higher education is also seeing a preference shift toward for-profit trade schools, like coding boot camps.

Even programs in the arts are becoming more real-world focused, trying to provide a supply of graduates to suit areas of demand. City University of New York, for example, offers three-semester masters' programs in journalism and film production that are taught by working professionals and are very focused on making job contacts and training people to produce content and video for online audiences.

Acquiring cutting-edge skills and getting specialized training is increasingly essential.

Traditional Processes Have Become Digital

I mentioned that I took a course in the then-new area of computer programming. The final exam involved coding a punch-card. Within a couple of years, punch-cards—and with them, my "expertise"—were no longer needed.

The first portable computer came out in 1981, the first laptop in 1991. Spread sheet and power point software became standard, and computer skills quickly became essential. In 1998, a very experienced editor friend of mine missed out on a job because she hadn't yet learned QuarkXPress, the 1990s software that had quickly become the industry standard for the print media. It had eclipsed PageMaker, the software program that had been introduced in 1985, and was itself made obsolete by the so-called "Quarkkiller" Adobe InDesign, which debuted in 1999. (When my friend had left her corporate job in the early 80s, everyone was still using electric typewriters.)

In her book *Now You See It*, Cathy N. Davidson, co-director of the annual MacArthur Foundation Digital Media and Learning Competitions, says that 65% of grade school students today will be doing work that has not yet been invented. She suggests that getting a four-year degree in any technology field makes little sense when the field will likely be radically different by the time the student graduates.

Whether your primary training has been completed or is still in progress, the path forward will certainly involve continuing education (CE). CE has long been customary—and obligatory—for people in many fields: medical sciences (physicians, nurses, dentists, acupuncturists), social sciences (professional counselors, social workers), legal services (lawyers and paralegals) and financial services (accountants, tax preparers and financial professionals).

Cosmetologists, aestheticians, and nail technicians are also obliged to take CE, as are certified members of the National Association of Professional Organizers and some architects, engineers, project managers, and safety personnel. (Requirements vary by state, licensing board, and certifying agencies' rules.)

One purpose of CE is to broaden expertise. For example, doctors serving upscale communities may benefit from expanding their knowledge about issues more common in poor, ethnically diverse urban areas, and generalists in rural private practices may need exposure to cutting-edge procedures and resources. People in any field can benefit from acquiring additional information. But continual technological change has made CE essential.

Competence-destroying transformations have wiped out entire business sectors, including video rentals and travel agencies. In the movie *Other People's Money*, Danny de Vito plays a corporate raider addressing stockholders unwilling to acknowledge the bleak prospects for a company whose product has been made obsolete by new technology (just as cars killed the demand for horse-drawn carriages). "At one time there must have been dozens of companies making buggy whips," he says. "And I'll bet the last company around must have been making the best damn buggy whip you ever saw. How would you like to have been a stockholder in that company?" he asks, before making his point: "This company is dead."

Traditional processes have recently been transformed into digital, mobile, and virtual ones. And ongoing, rapid introduction of new technology is necessary for companies to be economically efficient and globally competitive. Practitioners in every category and profession will need to increase what they know and can do on an ongoing basis.

Even in our personal lives, we have to keep learning just to deal with everyday needs. When your smartphone is upgraded to a new operating system, your new software may not be able to access your old computer files, and your DVDs can't be loaded into your new Apple computer. The only option is to adapt.

The digitalization of American life may be the quickest and most dramatic example yet of the adoption of what economists call "a general purpose technology" (GPT). Wikipedia defines this as a technology that can affect an entire national or global economy to drastically alter society through its economic and social impact. (Past examples include the steam engine, electricity, and interchangeable parts.)

Tech Skills Are Required of All Employees

According to a CareerBuilder survey, 69% of employers said every job is essentially a "tech job" because some part of the job requires the ability to utilize various technologies.

The 2017 *Brookings Report on Digitalization and the American Workforce* shows how dramatically digitalization transformed the nature of work.[134] The authors of the report quantified both the overall knowledge of computers and electronics required by a job plus the centrality of computers to the overall work activity of the occupation. Next, they defined jobs as digitally high level (score of over 60), mid level (33 to 60) and low level (below 33). Nearly 4 million of 13 million new US jobs created between 2010 and 2017—30% of them—required high-level digital skills, and 66% of new jobs required either high- or mid-level digital skills.

Some jobs accessible to new or less-skilled workers have now become medium or highly digital. And the required score level surged even in job categories where you might not have assumed much technological mastery would be needed, such as secretary, elementary school teacher, and registered nurse. For a construction manager it went from 17 to a whopping 60.

The job of human resource specialist, which had a digital score of 37 in 2002, got a score of 60 in 2016. For law, the digital score went from 34 to 58 in the same period, for financial management from 41 to 60. Other professionals similarly affected were chemical engineers, actuaries, and administrative support workers.

A 2017 study from Capital One Financial Corp reports that "some 82% of middle-skill jobs are now 'digitally intensive,' a 4% increase over the past two years." Even more important, it said that for an employee to advance, digital skills were necessary, calling them "a vital component if middle-skill workers are to remain middle class."[135]

Even the lowest-tech jobs like waitperson or doorman may require some facility with computers for functions like entering orders and recording deliveries, and people in those positions have to become competent and current with each update of the software and hardware.

To meet the demand, there are now accessible, low-cost (even free) programs to upgrade vital computer skills at local libraries, adult education programs, community centers, senior centers, colleges, and even computer stores themselves (if you purchase equipment). Job applicants can expand their opportunities by getting Microsoft Office Skills certification at various levels, for Word, Excel, and PowerPoint, via classes online or in person.

The comedian Jerry Seinfeld says that wardrobe evolution for most men stops when they get married. Active learning for many people stopped when they graduated from college. But society's needs are being modified and transformed.

New tools for artificial intelligence data (used to see what each student does or doesn't know and how each student learns best) and virtual and augmented reality (that provides new possibilities for learning) are changing our thinking about ways to get educated.

The idea of when to get educated is changing as well, and there are many new learning opportunities for people whose student days are well behind them.

Continual Learning Is the Way to Stay Marketable

"I believe the accelerations set loose by Silicon Valley in technology and digital globalization have created a world where every decent job demands more skill and, now, lifelong learning [to] add value beyond what machine and algorithms offer," Tom Friedman wrote in *The New York Times*.[136] He quoted education consultant Heather McGowan who says a "set dose" of education isn't enough. "The new killer skill set is an agile mind-set that values learning over knowing."[137] In today's world, we're destined to become rookies over and over again.

Helpfully, study options in distance learning are constantly expanding. By the end of 2018, over 101 million people from around the world had enrolled in Massive Open Online Courses (MOOCs), Coursera, EdX, and other online platforms, some affiliated with major institutions.[138] They offer an introduction to many fields of study without the long-term or financial commitments required of enrolling in a full bachelor's program. Kahn Academy, which provides free online learning tools that are used in elementary, middle, and high schools to augment teaching, also attracts adult learners who need remediation in areas like math before pursuing more advanced studies.

Of the early MOOC enrollees, 20% did not have an undergraduate degree and 80% did, 60% were employed full-time, and 60% came from developed countries—the 36 members of the Organization for Economic Co-operation and Development (OECD). But a Harvard Business School article in 2015 disputed the allegation that MOOCs were "merely a diversion for the well-educated and well off."[139] Of the 52,000 who responded to a survey sent to 780,000 people who had taken at least one course, 72% reported career benefits, and 61% reported educational benefits. A substantial proportion reported very tangible benefits, among them getting a new job, starting a business, or completing prerequisites for an academic program.

Many colleges and universities offer specialized training and certification in diverse areas. Harvard and MIT began offering online courses in 2012, and within four years some 290 course offerings had attracted 4.5 million participants and resulted in 250,000 certifications.

For a fee, many online programs offer verified certification for completing online study in diverse areas including global transactional banking, appraisal studies, legal interpretation, digital marketing, LEED accreditation (proficiency in "green," sustainable design, construction, and operations standards), certified public accountancy, and more. Some can be obtained in a year or less. All have the potential to augment present skills and lead to new job opportunities. (Enrolling and paying for a verified certificate makes it ten times as likely that you'll actually complete the course.)[140]

Concepts like Mozilla's "open badges" and Udacity's "nanodegrees" program can also enhance anyone's marketability. Mozilla uses open, shared technology from different sources, and learners can earn badges that testify to accomplishment, skill, or interest at any level for both defined job responsibilities and more abstract, people-oriented abilities. The badges can be displayed on résumés, websites, social networking profiles, job sites, and more. Udacity's "nanodegrees" are part of a project and skills-based educational credential program. Admission is available to everyone, and the programs are built by learning companies (though they are generally specific to the company rather than transferable). As of early 2019, there were 10,000 students enrolled in the MOOC-based degree program that offered 45 degrees in all, up from 15 the previous year.

Current online degree programs differ from earlier ones in several ways. They may be cheaper than on-campus equivalents; they may allow you to determine the size of your own course load; and the admissions requirements may be more flexible.[141]

Anyone of any age without a degree or daunted by the cost of time and money that a four-year program requires should also consider the many two-year programs offered at community colleges, especially because they are more likely to be focused on employment goals than four-year schools. A 50-year-old friend of mine who left college to raise kids decided to go to work once her kids were grown. She earned a two-year degree as a dental hygienist after her research indicated that the field offered the highest starting salary for the shortest amount of preparation time.

Continual learning is a means to professional growth that can help people of any age keep a present job, get a promotion, or open the door to a new career.

Do You Need Retraining or Just Repackaging?

Older workers may have allowed themselves to be pigeonholed by starting their career working in a particular area, staying there, and perhaps even advancing to a leadership position. However, in the changing global environment, the competencies acquired in that narrow niche may at some point have no marketable value.

Nevertheless, at any job you are likely to acquire certain proficiencies that can be applied elsewhere. That is why résumé specialists suggest that when you prepare your résumé, rather than simply listing prior titles and the particulars of your *duties* at past jobs—the so-called "hard skills"—you must give a fuller picture of yourself. You should find a way to describe transferable competencies that may be categorized as "soft skills."

"Hard skills" are specific: copy writing, coding, graphic design, data analysis, finance director, leasing consultant, and so on. "Soft skills" are abstract: self-motivation, time management, decisiveness, confidence, problem solving, team playing, curiosity, persistence, and (highly desirable) leadership. And they are more transferable, especially when you are moving from one type of job to another. For example, if you're moving from sales to work in customer service, "bidding a job" (which involves hard skills like making cost estimates and calculating profit margins) is far less important than "proficiency at client retention" (which requires "soft skills" like empathy and creativity).

Venture capitalist Morgan Housel suggests thinking in terms of what he calls "behaviors" that are another way to define your transferable skills. He says to think of what you are currently doing and think of a way you can use that ability in a different manner. By his definition, gathering statistics is a skill; finding facts is a behavior. Identifying smart people is a skill; managing smart people is a behavior. Being good at your job is a skill; working well with others is a behavior.

Housel uses as a case in point General Recorded Tape (GRT), a company that made 8-track tapes. Its merchandise was in 90% of music stories when it went public in 1968. But by 1979, when 8-tracks were superseded by cassettes, the company went bankrupt. He compares GRT to other companies that reinvented themselves, like Disney (that started in cartoons, moved on to sell theme park tickets and then to run cable channels) and IBM (that developed computers, sold computers, and then sold services that let you get rid of computers).

For an example of how this kind of reinvention works on a smaller scale, I was reminded of a profile I'd read about Richard Grace, whose company Grace Manufacturing fabricated printer parts. Etching steel printer bands with ferric chloride, a corrosive salt, produced bands with edges so razor sharp that they sometimes caused minor accidents. When the printers for which he manufactured parts became obsolete, Grace used one of his company's transferable skills—creating sharp edges—to develop a line of woodworking tools. Unexpectedly, it became very popular with home cooks for grating cheese. He now produces an entire line of kitchen tools, and, more recently, a medical product line.

Thinking creatively about what you offer can help you realize how you're marketable. Surprisingly, technology can give a boost there as well.

Robots May Help Identify You as a Job Prospect

Companies like Eightfold and Zip Recruiter now use machines that not only scan words on a page and match them to words in a job description but also use algorithms that can assess your capacities in more macro ways. If, for example, you are applying for a job at a company or in a field you're not familiar with, you may not know the jargon that will pique an employer's interest. Artificial intelligence (AI) can identify the skills and aptitudes your previous experience has given you, qualifications you may not have thought to list, and see how they fit with the qualities an employer is seeking; it can match what they think you're saying with what the company appears to be looking for.

AI knows, for example, that 90% of software engineers at the software company Intuit know Java, even if job applicants from Intuit fail to mention it on their résumés. It knows that certain disciplines (such as natural language processing at the University of Massachusetts Amherst) produce high-achieving workers, that chess players are likely to be good coders, and that basketball players make good salespeople. It knew that someone with a master's degree in statistics who also knew certain programming languages would be a good data scientist even though those keywords never appeared on his résumé.[142]

And while it's true that a majority of hiring managers surveyed currently consider a college degree the most reliable test of a candidate's skills, knowledge and perseverance, a 2018 study from the Northeastern University Center for the Future of Higher Education and Talent Strategy reported a growing deemphasis on degrees and more reliance on data and analytics to indicate skills and competencies.[143]

There is a potential downside here. A company accustomed to looking for candidates with MBA degrees might be able to find a similar combination of skills in people without degrees, which might depress salaries for those who have them. Or, if you were trying to identify what makes a great NBA player and studied NBA players only, you'd find no connection between performance and height because the whole universe you were studying would consist mostly of very tall people, so you might not factor in other important abilities, like speed and coordination.[144]

On balance, though, it seems that a digitized review of your résumé might open up possibilities to you as well as to employers. If you're job-hunting based on what you do now, a conventional jobs search engine might bring to your attention possible work as a marketer or software engineer. But with the technology that allows companies to model your behavior based on your browsing histories and interests, AI can infer what you like and your hidden skills and do "profile augmentation." After studying your search engine data, on-site search data browsing history and pages visited and shared content, AI can recommend jobs related to what you *want* to do. Its capability to match the person with a job will only grow.

AI can also open up jobs that currently don't exist or for which few people currently hold a title—like Blockchain Developer. It would know that many people with a background in software development could master the necessary skills required for that title.

Some predict that AI eventually may even be able to help revolutionize and customize education by creating a machine that can study you individually and, by identifying your strengths and weaknesses and discovering how you learn, will know how to teach you most effectively.

Opportunities for Older Workers Are Expanding

By 2022, 25% of the American workforce is estimated to be comprised of workers age 55 and older, according to the Bureau of Labor Statistics.[145]

Stressing the qualities that older workers offer—institutional knowledge, reliability, experience, and flexibility—no doubt helped AARP get General Mills, Boise Cascade, and Kimberly Clark to sign a pledge to support workers in diverse age groups in all kinds of jobs and a diversity of population. A number of companies, such as Walmart, Home Depot, and CVS have been hiring older people for jobs like stock checker, greeters and the like, and in summer 2019 McDonald's teamed with AARP to find seniors to fill 250,000 jobs.

However, when searching for new job possibilities at a management or professional level, mid-career workers may have found their age to be a negative factor when looking for employment. For 50 years, it has been illegal under the federal Age Discrimination in Employment Act, or ADEA, for employers to treat older workers differently than younger ones with only a few exceptions.

Nevertheless, in a paper published in 2016 by the Center of Retirement Studies, a researcher reported that while people over 50 become chief executives and sales engineers at twice the rate of people ages 30–49, in some categories the prospects for older people are definitely less bright.

Lawyers over 50 are 26% less likely to be hired than younger ones, and older computer scientists are 51% less so. An Urban Institute study noted that between 2008 and 2012, workers age 62 and older with a college degree had less than a 50% chance of finding work even after two years of actively looking; those without a degree had only a 35% chance.[146] (Surprisingly, women encounter ageism in hiring less often than men—because younger employees perceive older women as less of a threat to their own careers.)[147]

But things are changing.

When Barclays expanded its apprenticeship program for candidates older than 50, even those from an unrelated field, it stressed that it wasn't motivated by public relations purposes but had a genuine interest in finding people with life experience who could relate to people seeking auto or mortgage loans and eventually become good, full-time loan officers. (Barclay already had some tech-savvy older workers to help older customers with online banking.) By 2016, more than a third of those recruited were over 50.

The National Institute of Health, Stanley Consultants, and Michelin North America are among other institutions that have established programs designed to attract and keep workers past age 50. PwC, Regeneron, Harvard Business School, and McKinsey offer "returnships"—internships for older workers—although, like internships for students and recent grads, some don't pay, and they may distract from or postpone a focused quest for a particular job.

What will certainly open new opportunities and brighten job prospects for seniors who would prefer to work at an interesting position rather than retire are some important demographic realities.

- The group is growing in size and thus in power. The Census Bureau reports that by 2020, for the first time in human history, people aged 65 and over will outnumber children under age 5, and the two age groups will then continue to grow in opposite directions. To gain the goodwill of this growing and presumably increasingly influential segment of the population, some futurists predict companies will not only assist employees with their current careers but also prepare them for new opportunities as they age and provide supportive services for older employees much as they do now for female and minority workers.
- Today's seniors are not only better educated and healthier than yesterday's, but also, in an economy dominated by service, capable of working longer in retirement.
- Medicare continues to cover insurance for seniors. Employers faced with a continued increase in benefits expenses will look even more favorably on hiring them.
- A study from Georgetown's Center on Education and the Workforce suggests that judgment/decision-making, communications, analysis and administration will be the four most in-demand competencies in the labor market—the competencies that experienced workers are more likely to have developed.
- Perhaps most critically, there's going to be a major labor shortage. The Georgetown Study estimated that due to the combined effects of a falling fertility rate, tougher immigration, and the number of people who choose to retire, the United States would fall short by

> 5 million workers with a post-secondary education by 2020. About 65% of the vacancies would require some post-secondary training—either an associate degree (7 million jobs), a baccalaureate (23 million) or a graduate degree (6 million). Some 5 million would require some certification (in fields like child care or IT) and 10 million would require at least some college.

Aware of some of these changes, many seniors have been inspired to enroll in specialized vocational and certificate programs. While the average age at coding boot camps is just under 30, there are one million students aged 55 or older (out of a total of 45 million students globally) at Codecademy, a MOOC course that offers training in coding and other new skills. The courses are mostly free, but there is also a "pro" option that provides more customized lessons and tutorials as well as live tutoring.

An increasing number of other interesting resources and possibilities are also available to older people seeking challenging work. Among them:

- AARP's Search in the Digital Age offers assistance to those 50-plus who are either looking for employment or hoping to switch careers, helping them navigate through job sites and marketing themselves so as to secure an interview.

- Investment bankers, financial analysts, corporate attorneys, and others interested in offering their skills to organizations where they are needed can apply to the nonprofit Encore.org for one of its one-year fellowships, which are typically at another nonprofit. Although they are temporary and pay just $25,000, they are designed to be a springboard to help mature workers reenter the job market.

- Adventurous seniors looking for something different should check out Coolworks.com, which seeks "older and bolder" employees who want challenging seasonal or temporary work such as living and working in a National Park or training to be a raft guide.

Adaptability Has Always Been the Key to Survival

New technology has always caused some job displacement, but there is reason to be optimistic about job expansion. According to the World Economic Forum, by 2022, the shifting labor division between humans, machines, and algorithms will result in 75 million lost job roles, but an addition of 133 million job roles will result in a net gain of 58 million. The rapid emergence of new jobs has left a skills gap that is projected to leave 2.4 million positions unfilled between 2018 and 2028. Digital skills like programming and technology will have to be combined with human skills like critical thinking, emotional intelligence, and creativity.

Thanks to technology, a lifetime of work will involve continual change—change of employers, change of career, change of home base. By evolutionary design, adapting to change isn't meant to be easy. It's built into our DNA that our first reaction to something new is to back off and the next is to determine whether it's something to be feared. It's not change itself that bothers us but the fears that change raises: *Is there some sinister cause behind the change? Will it serve a purpose? Will it actually work? Will I be able to adapt? Will it benefit or harm me? Will I lose my identity, my financial security, or things that are familiar to me?*

The way to quell those fears is by planning and, rather than regretting what is lost, finding opportunities in what is coming.

8

Protecting Your Cyberself

Guard Your Identity and Your Legacy

Your lifetime goal:
Keep the wrong people from accessing your cyberself.
Your legacy goal: Ensure that the right people can.

As of April 2019, 56.1% of the world's population had access to the internet where they could animate a virtual self.[148] Creating that vast number of cyberselves is arguably technology's most awesome achievement.

Especially in developed countries, where 81% of the population has internet access, cyberselves have extraordinary powers and responsibilities: password access to your phones, voicemail, and computers; to lockboxes, banks, brokerage, credit cards, and shopping accounts; to domain names, web pages, social networks and photo albums; to Google Drive and other online storage, subscriptions, and entertainment that you connect to.

Cyberyou has intimate knowledge of your possessions and your obsessions, what you own and who you are. It manages your finances; updates your personal data in multiple locations; curates your photo collections; oversees your intellectual property holdings including online games, licensed software, books, music, and videos; broadcasts your opinions via blogs and reviews, and distributes your images via photos and videos.

We are very protective of our corporeal selves, carrying umbrellas and buckling into seat belts, tucking wallets into pockets and holding tight to handbags. We equip our homes with security devices. We stay alert to our surroundings, monitor our health, make prudent choices about where we go, with whom we associate, and what we do.

We tend to be far less vigilant about our cyberself, despite the depth of its powers and the breadth of its activity. By 2020, it is estimated the average internet user will have over 200 separate accounts of various types,[149] coupled with the fact that it often talks to strangers and travels to shady places where stolen identities, hacking, and malware abound.

With our knowledge and permission, our cyberself reveals huge amounts of information, often quite deliberately, sharing our intimate thoughts and images, filling out forms, and registering credit cards and medical and other information on websites. It may also put us at risk without our full awareness or consent, by allowing our behavior to be tracked and algorithms to profile us as it surfs among sites and networks on social media.

Whenever you respond to an email cleverly disguised to look like a legitimate inquiry from the bank, log into a Wi-Fi hotspot that's not secure, or send confidential information to an online advisory service, you put your money and even your identity in danger.

Identity Fraud Is Down, But Credit Card Theft Is Up

Many of us who have had the experience of canceling and replacing a stolen credit card due to fraudulent charges might be surprised to hear that identity fraud declined nearly 15% from 2017 to 2018 when it affected just an estimated 5.6% of US adults—still an unfortunately large number of people, though a greatly reduced figure. This was almost entirely thanks to the spread of chip card technology. The old magnetic strips were encoded with your credit card number, expiration date, and other information, whereas in the chip there is a microprocessor that generates a unique code for every transaction. Information hacked from a chip is useless for subsequent transactions.[150]

And now there's a highly effective new tool to thwart would-be thieves who've turned their focus away from brick-and-mortar merchants to online retailers: virtual account numbers that card companies such as Bank of America, CitiBank, and Capital One have begun to issue to online banking customers. You set the parameters for their use—one time only for an online purchase, or limited to use with a particular retailer for any online purchases or during a specific time period. Both the retailer and the fraudster would get access only to your virtual number, not your credit card number.[151]

Still, the 2019 edition of the annual *Identity Fraud Study* from Javelin Strategy & Research reported that identity theft (especially of mobile phones to get access to financial data) and new-account fraud (opening accounts using someone else's identity) remain "alarmingly common" and at levels much higher than in the past. Fraudsters have become better at meeting authentication challenges. And victims have been increasingly faced with shouldering the financial consequences.[152]

The Consumer Sentinel Network, maintained by the Federal Trade Commission (FTC), which tracks consumer fraud and identity theft complaints that have been filed with federal, state, and local law enforcement agencies and private organizations, said that in 2018 consumers reported losing about $1.48 billion related to complaints about impostors and scammers, an increase of $406 million over 2017.

The Network also reported that identity theft represented 15% of all complaints. Though for several years the rate of problems in this category had been falling, they began to increase again in 2018 when they rose 19.8% from 2017.[153] What's particularly troubling is that children are 50 times more likely to be victimized than adults—about 1.3 million of them each year, according to The Identity Theft Resource Center—and the consequences may not come to light for a long time. Only when denied a student loan or a first mortgage, young adults might discover that their identities had been stolen many years ago, and since then someone has been abusing and destroying their credit.

The consequences of identity theft, even if not directly financial, may have a negative effect on one's personal and career prospects. In 2015, hackers got into the account of a company that introduced people to one another for the purpose of extramarital affairs. They then posted online not only users' names but also their email addresses, profile descriptions, postal addresses, GPS locations, and personal data, as well as a separate file containing credit card transaction data (although without full payment information, card numbers, or billing addresses).

Even if we take reasonable measures to protect ourselves against identity fraud and identity theft, motivated parties can succeed in doing damage. While technology has found ways to curb some efforts at fraud and theft, it has also given new tools to the unscrupulous. For example, in a form of fraud called "morphing," they merge an online photo of you with a photo of a thief and create fake supporting documents to get a passport in your name. The process is another variant of the "deep fake" videos that appear on the internet.

But not all cyberthreats come from outsiders. Sometimes we ourselves are to blame for being taken advantage of.

Cyberyou May Forfeit Your Rights

The Electronic Signatures in Global and National Commerce Act (ESIGN) passed in 2000 made it legally acceptable to use electronic records and signatures for a wide range of purposes. Most companies rushed to use Cloud services both to stay competitive (customers and clients like the convenience) and because signing agreements online is cost-effective as well as fast. Also, since papers never leave the building, and electronic paperwork is harder to lose, some believe online contract-signing is more secure—although recent hacking successes and your own experience with misfiling a document because of a wrong keystroke or losing data because of some technical error may have made you a bit skeptical about the claims of electronic infallibility.

Signing contracts online may make you vulnerable in ways you aren't aware of. As I have described, while the European Union and some states in the United States have instituted regulations that restrict institutions from gathering private data, our cyberselves routinely relinquish privacy rights in signing TOSA agreements. But in signing those agreements, they may be ceding other rights and assuming other obligations with financial consequences that your corporeal self might be more concerned about.

One Deloitte survey of 20,000 people found that 91% of them signed online agreements without reading them—and among people ages 18 to 34, the rate was a whopping 97%. Other studies confirm that over 90% of consumers accept legal terms and conditions without reading them.[154]

A couple of researchers wanted to see to what extent people would put down their guard when presented with online contracts. They created a fake social network site with an accompanying terms and services agreement. It stipulated not only that anything users shared would be passed along to the National Security Agency *but also that they agreed to give up their first-born child as payment*—though that didn't stop 98% of participants from signing it. (Even though a parent might be tempted at one time or another by the second part of the deal, the compliance rate seems astonishingly high.)[155]

Recognizing that customers rarely if ever take the time to look over the Terms of Service Agreement, companies now may not even bother to provide them in their entirety. I rented a car recently and was handed a small electronic pad that they wanted me to sign—without my being able to look at even a portion of the actual document. It's disconcerting that I had only to create some kind of indecipherable squiggle on a tablet to commit myself to whatever the automobile company wanted to hold me to.

As humorist John Oliver puts it: "The great truth of America: if you want to do something evil, put it inside something boring. Apple could put the whole text of *Mein Kampf* inside the iTunes user agreement, and you'd just go, 'Agree. Agree. Agree. What? Agree. Agree.'"

Though companies generally use these terms of agreement only to protect themselves from legal problems, some are much more aggressive about claiming rights. Consider this: Instagram's TOSA says that you own your own photos but also says, "Once you have shared User Content or made it public, that User Content may be re-shared by others." One artist made copies of some Instagram photos and sold them. He didn't license them from Instagram, which would have given the owner some recourse against Instagram; he just read the TOSA to mean anyone, not just Instagram, could reshare material.

While that agreement was particularly egregious, there are other, similar examples. A presentation software company acquired very broad rights to use and exploit the user's material without any obligation, and another firm's contract secured its right to have a user cover the cost should it have to sue him or her for late payments.[156] (One anonymous cynic has written "Terms of Service agreements are the universal solvent for intellectual property rights.")

Of course, rather than sign the TOSA, you may choose to do without the product or service in question, but most of the time, that's an impractical alternative.

There are many other threats to cyberyou. Hackers can blackmail it, hold it hostage, manipulate it or incapacitate it, transfer assets, and manipulate accounts. Intellectual assets that cyberyou puts online are easily stolen, now that technology has made it so easy to copy text, sound, and pictures.

The very existence of cyberyou is precarious. A malevolent party may use stolen data that looks legitimate to get into your system with a new type of "wiper virus" that causes it to disappear. Of course, a simple technical glitch may obliterate it as well.

And consider this: If you don't plan carefully, long after your corporeal self is gone, your cyberself may stay in control of assets your heirs cannot access and may not even be aware of.

Threats to a Legacy That's Neither Here Nor There

After Anthony Bourdain died, newspapers reported that he had bequeathed to his estranged wife his household effects, including his cars, furniture, books—and frequent flier miles. The globe-trotting chef, author, and TV personality had probably accumulated an impressive number.

The record for accumulated miles at the United Airlines program went to the owner of an automotive sales and dealership management consulting company, who in 2018 had accumulated more than 19 million of them. But far less well-traveled folks who are motivated to accumulate points and follow tips from websites like thepointsguy.com and millionmilesecrets.com can rack up numbers that translate into significant assets. One industry expert calculates that, on average, half a million miles may be worth between $4,000 and $10,000—or more. For just 110,000 miles, one enterprising credit card reward player scored an ANA airlines flight from Los Angeles to Tokyo with a $20,000 price tag.

Airline mileage is just one of the items that are in a grey area between tangible and intangible assets. As such, they may be overlooked when people are preparing their wills—another of the issues technology has created. As recently as the 1980s (and in the case of many estates, perhaps long afterward), an executor could deal with the holdings of a deceased person in a rather straightforward way by cleaning out the house, going through file cabinets, and collecting all the incoming mail for three months or so. By then the executor would probably have enough information to know what obligations remained, where the assets were and what they amounted to. But the assets in cyberspace are quite literally neither here nor there.

Since many people have relinquished so many tasks to their cyberselves, an executor now must investigate how the decedent organized finances and assets online. There are several areas of concern:

- *Data management*: documents in digital formats on servers scattered across the globe, online, and in the email inbox (a catchall for paperless receipts and credit card statements, insurance information and other financial transactions).
- *Archival responsibilities*: control of photo and video-sharing sites as well as status updates, blog posts, and musings left behind at social media sites like Facebook, Twitter, and LinkedIn.
- *Asset control*: access to the bank, brokerage firm, cryptocurrency and credit card company accounts, direction of automatic payments for utilities, mortgage and other ongoing obligations, and oversight of account activity records from Amazon, PayPal, eBay, and other vendors.

Together with the increase in the exemption that makes estate tax likely only in the case of very high-net-worth individuals, the proliferation of digital accounts and the sheer amount of information that is accessible only online has changed the nature of traditional estate planning. Planners today are not only focused on estate tax issues but also increasingly on digital asset management—making certain that access to online information has been provided, and, in addition, that the correct legal documents have been prepared that dictate how various digital assets should be disposed of and who is responsible for doing so.

Without this kind of oversight, on your death (or your incapacity) all or some of your assets may be subject to hacking and fraud or even lost entirely if your heirs can't easily—or perhaps ever—locate or gain access to them or because they are unaware that they exist.

Even before consulting with an estate lawyer, you should create a digital will to give to both the lawyer and someone you designate as your digital executor. A digital will is an informational rather than a formal document that gives family members an inventory of your online presence and provides the lawyer drafting your will with some essential information. It should include the name of the person you want to designate as your digital executor, plus an alternate. (A selected list of what to include in your digital will is at the end of this chapter.)

Also include instructions for what should be done with each asset. You may want to delete your Facebook account, for example, but share your photos.

Ensuring Access to Your Digital Assets

There are two particular areas of key concern with your digital legacy. Someone in charge of settling the affairs of your estate may see an incoming email about an online financial account that requires a response or a notice that a payment is due. Nothing can be done without the user name and password to get into the account. But having the passwords may not be enough to allow someone other than you access to your account. Federal and state laws are confusing and sometimes contradictory with respect to the transfer of accounts to heirs.

By the terms of the 1986 Stored Communications Act, accessing a person's online accounts and digital property could be a crime under federal or state law. Together with the Computer Fraud and Abuse Act, the law was created to prevent the unauthorized access of computer systems and certain types of protected data while the owner of the digital asset was alive. However, it did not provide any provisions in the event of death or incapacity.

Besides, just having access to an account didn't necessarily give a named fiduciary the legal authority to take action, especially if the Terms of Service Agreement from the supplier (the email server, for example) didn't permit a transfer of ownership. Heirs who tried to access an account could be found guilty of hacking—even if the owner's dying wish was to grant them permission. In response, the Uniform Law Commission (ULC) successfully established the Revised Uniform Fiduciary Access to Digital Assets Act (RUFADAA) of 2019. RUFADAA was enacted in thirty-four of the fifty States.

On one hand, it is meant to give someone's personal representatives, guardians, agents acting under powers of attorney, and trustees the legal authority to manage digital assets and electronic communications including computer files, web domains and virtual currency, just as they are allowed to manage tangible assets.

On the other hand, it was meant to provide services such as Yahoo, Facebook, and Google (the custodians of digital assets and electronic communications) the legal authority they need to interact with fiduciaries of their users while still abiding by the user's privacy expectations. Under the Act, a Terms of Service Agreement controls your digital assets to protect your privacy, *unless* you have given explicit instructions otherwise, by specifying what to do with digital asset rights in a legal document such as a will or power of attorney or using an online tool.

One such tool is Google's "Inactive Account Manager." The person who opens the account can determine how to handle an account that's inactive for whatever specific time period the owner wants to set. After that date, Google sends an alert. If the account holder fails to respond to the alert in a timely manner, Google will notify a friend or family member—or the digital assets manager you have named—to confirm your death before carrying out your instructions.

The list below shows how different platforms handle legacy issues. (Note: These tend to be revised periodically, so you should check to see that you have the most recent versions when you do your estate planning.)

- PayPal lets an estate's executor close out the account and will liquidate funds via a check to the estate.

- Twitter allows family members to deactivate a deceased member's account.
- eBay asks that you contact its support team.
- Yahoo will help you close an account but says neither the account nor any of the content is transferable.
- Facebook lets family members convert an account to "memorial status." When it receives proof of death, it deletes sensitive personal information and changes the status of the account.

As a general rule, the Terms of Service Agreement will favor the online providers and block access to anyone who isn't the account owner. So be proactive in specifying your wishes, or you may leave your heirs with a potentially costly, long-running, unresolved headache.

For accounts or websites attached to a business, you may need to name an employee to act as digital executor, though in some cases any family member or friend who knows how to access your account can act for you. (That person should know if any of your accounts are automated.)

Obviously, the disposition of some assets with online records (bank accounts and currency accounts) would be covered in a traditional will, as would monetized online assets such as a website or online business. So can a domain name that has financial value. The worth of some can be significant. If you have such an asset, mention it in your will. A Twitter account that has a significant number of followers may also have enough value to be sold. If you believe that is the case, be sure to leave explicit legacy instructions about how to handle the sale and proceeds.

An artist, writer, or composer may have intellectual property to protect, not only completed works but unfinished projects on a computer or hard drive or in the Cloud. Intellectual property is especially vulnerable to theft, not only by foreign nations who don't honor international copyright agreements and steal property on a grand scale, but also because it is simple to copy written materials, videos, audio tracks, and art works on paper. Soon, as the technology improves, it will no doubt also be easy to copy three-dimensional works, from oil paintings to sculptures. Creative people must designate a personal representative to monitor and handle those aspects of his or her legacy.

Passing on Digital Currency Is a Separate Concern

Bitcoin is a digital payment network run by computers around the world where bitcoin currency—which is digital and can be sent electronically—is stored and transferred. Bitcoin tokens do not come in set amounts. Instead, the value is calculated by transactions that computers track. Similar to stocks, a bitcoin changes in value based on bidding, and the value can be calculated to eight digits. Owners can "store" Bitcoin currency in a "wallet" that has a public address. The bitcoins may be sent to and from the wallet when the owner uses a private key to authorize the transfer. That key is the only identifier of ownership to a wallet, and lost keys cannot be recovered. Some people put them on a thumb drive or on a piece of paper in a safe deposit box for safekeeping.

On your death, your bitcoins become inaccessible if you fail to leave a private key and/or an explanation of how to access it. Unaccessed Bitcoin currency simply disappears from circulation, which actually benefits other Bitcoin investors. Since there is a cap on the number of bitcoins that will be issued, having fewer available raises the value of the others.

Some choose to have their digital currency handled by an exchange, a company that offers to trade local currency for digital currency and vice versa. When your digital currency is held in an exchange, you don't have access to your private key, but use master passwords (which you set yourself) to authorize the exchange to make a transaction. You can reset your password, and your account is insured against hacks and loss of private keys. The exchange is technically less secure, since it is a third party between you and your wallet; and it's not really anonymous, since the website knows who you are and/or can monitor your capital gain in accordance with law.

In a terrible turn of events in 2019, one exchange was unable to retrieve digital currency and return cash owed to its customers (a total of more than $190 million) when its young manager died and no one was able to find the passwords.

Ownership of Licensed Property Is Unclear

When you die, what happens to licensed property—your extensive Kindle library or iTunes account—is another gray area.

Software becomes more and more embedded into everyday products, and manufacturers abetted by Congress have used code restrictions and copyright law, generally successfully, to create a world in which everything is licensed and nothing is owned.

One notable exception: Not long ago carriers could refuse to unlock the cell phone you bought, so you were stuck with the carrier you bought it from. However, in 2015 The Cellular Telecommunications Industry Association (CTIA) adopted new, less stringent policies. They allowed a phone to be unlocked once it was paid off or at your request no less than one year after initial activation, consistent with reasonable time, payment, or usage requirements.

But licensing has become more and more the model. Microsoft, for example, used to sell a current version of its software and issue regular updates for a small fee. You could continue to use the software as long as you wished. Now the company does sell a current version, but it is expensive, and customers are steered toward the licensing model, with a monthly payment. And although Kindle urges you to "Buy Now," you're not actually buying the books you order from this platform. Instead, you're licensing them, which Kindle's Terms of Service Agreement makes very clear. A fundamental rule continues to prevail about anything licensed, including digital books, movies and music: If you license it, you don't own it.

That rule was backed up in a court case. Capital Records sued to block Redigi, a startup, from proceeding with a plan that allowed people to sell "used" music from their digital libraries. The intended seller would download Redigi's software to verify ownership of the files, and then Redigi would remove them from the seller's files and move them to the secondhand purchaser, who would pay less than the original purchaser. In its defense, Redigi offered the "first sale" legal argument: that an original publisher or issuer can dictate the terms of the first sale but not resales. The judge ruled that this didn't apply because transferring digital media equates to making a digital copy—*whether or not the first copy is deleted*—and that is unlawful.

Apple and Amazon subsequently filed patents regarding digital rights management to further impede transfer of licensed material.

It's also not clear what happens to your digital book, movie, and music library if the source you bought them from goes out of business. In its 80-page document outlining terms and conditions for purchases from iTunes and its App Store, Apple does not say what would happen if it ceased to exist and whether you yourself (never mind your heirs) would even retain the right to use the music.

The cases of online companies that recently disappeared, such as Wii Shop Channel and the Ultraviolet movie streaming service, aren't encouraging. The Wii Shop Channel sold digital copies of video games, and when it closed down, you could download them to your Wii but not to newer Nintendo consoles. The Ultraviolet service was mostly a movie streaming service but did let you purchase movies; when it closed, it suggested transferring licenses to a competitor's service.[157]

Rewards Might Disappear Into the Ether

As already noted, be sure to mention in your will the existence of rewards points, mileage, and similar items, and state to whom they are being left, so the person can make the effort to redeem them. If you're careless about keeping records of rewards owed to you by credit card or frequent flyer programs, search online for "tools to track your rewards" to find an app that will help you out.

But leaving this sort of asset to heirs is not as simple. Experts in property law have generally reported that legally, rewards aren't considered the property of the rewards member and can't be passed along.

Still, having the user name and password may let you access the rewards at some companies. Southwest, for example, allows points to be used as many as 24 months after your last activity, and heirs who can access the account can use them or transfer them for a very modest fee. (With some airlines, the transfer fee is so large that claiming the reward is of little value.)

Even airlines that have strict written policies that seemingly prohibit transferring reward miles may be far more flexible in practice. Although American Airlines, for example, says in its program's terms and conditions that mileage, award tickets, and upgrades are not transferable upon death, it adds that at its sole discretion the airline may decide to pass them on, pending receipt of satisfactory documentation and after your heir signs certain forms.

Virtual privacy and legacy issues have very real consequences. It has been speculated that within the next decade or two, everyone will be assigned a unique digital identity that will be impervious to theft. But until that time, your best remedy is to recognize where your vulnerabilities are and take protective measures.

NOTE: The world of technology is constantly evolving. The following information is accurate as we go to press, but some of the sites may have changed names, become irrelevant or obsolete, disappeared, or in some other way become unreliable. It's my hope that you'll get some good ideas and use what follows as the jumping-off point to do your own research.

Strategies to Protect Your Online Privacy and Identity

Take Common Sense Measures

- Have a good antivirus program on your computer.
- Don't download software of dubious origins.
- Keep sharing of private information to a minimum. Use SSL protocols on your business or personal website. SSL certificates will protect the sensitive data transmitted from and to your website (like login details, sign-ups, addresses and payment or personal information) from being misused by attackers.

Freeze Your Credit

- Having a credit freeze with Equifax, Experian, and TransUnion prevents most companies from viewing your credit record unless you are already doing business with them. A thief can't get a credit card in your name because a company won't open an account without being able to check your credit.
- Freeze your kids' credit, too. It's the best way to protect them from identity theft. Since there's usually no negative information in their files, children make good

targets for thieves; and since most people don't take the precaution of freezing their kids' files, thieves may be able to operate undetected. There are instructions for how to freeze your kids' credit on the credit bureau websites. (And since in creditland, you're deemed an adult at age 16, try for an adult freeze for a child that age; but if you're unsuccessful, try the child route. Warning: a lot of photocopying is necessary.

- Major warning: Don't misplace the PIN number the bureaus give you, as you will need it when temporarily or permanently thawing your file or the one you've frozen for your children.

Use a VPN *(See page 119)*

Remove Cookies

- You can delete the cookies manually or use apps to do it. Some will delete all the data at once or allow you to choose what exactly you want to do: delete browsing history, delete download history, delete saved passwords, empty the cache, and so on. If you need help, do an internet search to find *the most recent* article about this subject in a source such as *PC Magazine* to get current advice. (You may even find instructions for dealing with specific browsers.)
- If you really want to browse and download anonymously, try using Tor (this stands for The Onion Router), free software developed originally by the Navy and now a nonprofit organization that hides your internet footprint. You can find a guide to what it is and how it works online. But it doesn't provide 100% security. You might want to use it together with a VPN browser.

Check What You're Agreeing to

- There are online sites such as Terms of Service, Didn't Read (tosdr.org) and TOSBack (tosBack.org) where you can find plain language versions of the online contracts you're supposed to sign.

Be Proactive About Privacy Rights

- Facebook is only one of many companies that have been revealed to violate privacy rights. In summer of 2019, both it and YouTube were accordingly slapped with large fines.
- Uninstalling and deleting your links to the most obvious perpetrators is not the solution. Letting companies of any size know that you care about your privacy and will walk away if they violate it can carry weight, especially if many people do it.
- Stay alert to new legislation that is passed to curb privacy violation. Work to support those efforts when possible.

Strategies to Protect Your Digital Legacy

Leave a digital will that is not part of the will your lawyer drafts. Wills do not change frequently, but digital information does (which is why you should update your digital will periodically).

The digital will should include passwords to all your devices:

- Computers
- Phone
- Tablet
- E-reader

- Camera
- Memory Cards
- CDs
- Cloud storage
- Hard drives

It should also include means to access the following:

Basic Information

- Documents
- Email, work and personal
- Blogs
- Photos and videos
- Social media accounts (Instagram, YouTube, Facebook, Twitter, Snapchat, LinkedIn, etc.)

Financial Information

- Bank accounts
- Tax records
- Credit card accounts
- PayPal account
- Venmo (and similar)
- Bitcoin (and other currencies)
- Investments
- IRS filings
- Insurance accounts
- E-File sites

Recurring Obligations

- Loan payments
- Phone bills
- Utility bills (gas, electric, water, sewage)
- Automatic e-payments
- Local property tax account

Home Management

- Apps and smart devices related to your home network, security, air conditioning and appliances

Shopping

- Amazon, eBay, etc.
- Virtual Assets
- Music
- Videos
- E-books
- Spreadsheets

Miscellaneous

- Rewards from airlines, credit cards, cashback offers
- Business points, discounts, vouchers
- Entertainment: Netflix, Hulu, etc.

9

Avoiding One-Click Solutions
Beware of Complexity Masquerading as Simplicity

Technology is a useful servant but a dangerous master.

Christian Louis Lange,
Nobel lecturer, historian, teacher, political scientist

By democratizing the process of investing and financial planning, technology has given you tremendous new capacities to build your nest egg. These days, making transactions is remarkably easier, speedier and less expensive than ever, and all investors have access to a wealth of information that was formerly available only to professionals.

Some people are so suspicious of the power of technology that they have shied away from investing because they think the game is rigged, and they believe it's useless to try to move as fast as the algorithms doing the trading. Many more have decided that investing and planning can be managed by allowing computers to make decisions. They're both wrong. Technology has given us some great new tools, but like all tools, to get the desired results—in this case, preserving your lifestyle for a lifetime—they must be employed together with human insight, caution, focus and deliberation.

Investing *Isn't* Another Computer Game

You've most likely seen many advertisements trying to persuade you to show up at your local Holiday Inn for a course that will make you a stock trader successful enough to put money worries behind you now and forever. They offer an easy system to follow or trend-spotting strategy to adopt. The most important thing you are likely to learn is something you should have learned before you signed up. There is no get-rich-quick path to investing or Da Vinci code that an instructor can help you crack. (If there were, he'd be enjoying a holiday trip instead of lecturing at the Holiday Inn.)

If you happen to be trading in a period of time favorable to a particular program you've been taught to follow, one that's perhaps great in a flat market but blows up on big swings or vice versa, you may start hitting a few singles with small trades and netting $100 here and $200 there. The danger of this is that it motivates you to start making bigger bets until sooner or later you may lose a bundle.

Not long ago, you routinely consulted a financial professional qualified with an advanced degree, a license, or both, and most likely a track record to help you put together a portfolio to meet your needs. How did so many people come around to thinking that such training and experience could be streamlined and simplified and packaged into a class that within several hours would make you equally adept at making investment decisions? The biggest danger, as with anything you attempt, is that you don't know what you don't know.

Doing the homework to put together a portfolio is something you can do yourself. As they say, it's not brain surgery. On the other hand, it's not a bathtub installation or couch upholstering, and you probably would not attempt those either, even though much less is at risk. Your real goal is to acquire an appropriate portfolio and an appropriate plan that is customized for your needs and, most important, to anticipate and make logical inferences about how future events are going to affect you. Assuming that making one-click investment decisions and following generic strategies is the way to succeed is like assuming that playing the kazoo qualifies you to join an orchestra.

Managing your money doesn't have to be a day job, but neither should it be a hobby. It takes real commitment to be effective and consistent. If you'd much rather be doing something else or if you're likely to get distracted, it would be best to seek some professional counsel.

The Popularization of Stock Market Investing

Not very long ago, only a small subset of Americans had the resources and inclination to invest in the market. At the time of the 1929 crash, less than 1% of the population held stock, and the stories of people totally devastated by that event seems to have scared off potential investors for decades.

Also, investment information wasn't widely available, high commissions on trades made even buying stocks an expensive proposition, and investment choices were limited. As a consequence, by 1952 only 4.2% of the US population, about 5.4 million people, owned stocks. But in 1954, the New York Stock Exchange introduced an investment program in which people could contribute as little as $40 a month. Other monthly investment plans were introduced by mutual funds. They grew even more popular when people began to purchase them for individual retirement accounts (IRAs), which were first created in 1974, and more popular still as a component of the 401(k) plans created by the Revenue Act of 1978.

Even as late as 1980, just 13% of Americans owned stocks, but eventually, people began to realize how effectively money could grow when properly invested. The number of American stockholders had reached a peak of 65% by 2007.

When I bought my own first stock in 1980, and for quite a few years afterward, the service of a stockbroker was required. Brokers were gatekeepers: They did the trades, but they also provided counsel. And, of course, of great value to private investors was that they had access to research tools that individuals did not, such as Moody's, ValueLine, S&P, and similar costly, subscription-only periodicals that were the primary sources of financial information in those days.

Brokerage firms had to bear the expense of supporting the library and the research staff, and everything was very labor-intensive. New monthly pages would be distributed by the handful, and staff people had the job of taking them to the library, pulling out the correct binder, opening the rings, removing the outdated report and inserting the latest one. The same process applied with updating regulatory changes.

When I became a broker myself, I would recommend a stock only after I had done some homework, was convinced that the quality of the company merited an investment, and understood its role in my client's portfolio. Now, as then, when clients ask me to get information about a stock they've heard is promising, I proceed in the same way.

As more and more people entered the market, and the investment choices became ever greater, people were still investing in the traditional way. They took comfort in the fact that they (often with the additional counsel of a trusted advisor) knew a lot about the investments they were making.

By the 80s, technology had begun to make significant reductions in the costs of printing, mailing, delivery and organization. Today, professionals on Wall Street can easily and quickly do their research on a computer and send it to a client for review and discussion. And while subscriptions to some of the financial reports are still too big an expense for most investors, a great deal of information is now widely available online. For example, while in the past you could get a company's annual filings only from your brokerage firm or by calling the company, they're now available online. You can access it at https://www.sec.gov/edgar/searchedgar/companysearch.html. For instructions on using the site, go to https://www.sec.gov/edgar/quickedgar.html.

Technology helped investors by bringing down the costs of the process enough so they can operate on their own, but it also eliminated some of the benefits. They lost the "protection" that a research department could provide along with the educational background, licensing qualifications, and experience that gave professional advisors an analytic filter through which they could consider an investor's purchases. Whether the filter was always high quality is another conversation. The point is that someone was there to poke at the information, inspect it, weigh it, and generally play devil's advocate.

Through their own homework combined with the advice of the professional they were working with, investors understood how product cycles, the economic environment, and the quality of management would be likely to affect the investment they made. That knowledge helped them stick with the companies they had selected even through rough patches.

In just the past few decades, such informed decision-making has become increasingly rare. Rather than investigating the merits of companies and investing in ones that met their criteria, many have chosen to let computers make their stock selections for them. And in deciding when to get in and out, they are guided by their feelings about the investment climate rather than conviction about their holdings. This can be attributed to a decline in the centuries-old practices of investing.

Shareholders Used to Be More Invested In Their Investments

The underlying concept of the stock market is simple: Sell shares in an enterprise that involves some risk but has a big payback if successful. It all began in the 1600s. Dutch, British, and French companies sent ships to India to fetch goods to sell back home. Because bad weather, pirates, and primitive navigation tools posed dangers that made the outcome uncertain, trading companies sought backers to fund the costs with the promise of sharing profits. We can assume those early investors understood a great deal about shipping practices. If they knew which routes were most vulnerable to pirate attack, when the weather would be most favorable, and which captains were the best sailors, they could calculate which voyages could have the best shot at arriving home safely.

For the next four centuries, the concept worked for other businesses as well. Companies learned to raise money to spread their risk and let others participate in their growth. At first, potential investors were most likely to be approached directly and, like the maritime investors, were probably familiar with the businesses they were asked to back. But eventually, as the number and diversity of businesses grew, stock brokers served as middlemen to present increasing numbers of opportunities to investors.

Some 4,397 firms were listed on US domestic exchanges at the end of 2018, down significantly from more than 8,000 during the dot.com peak in 1996.[158] Companies frequently disappear: They become obsolete, go out of business, or are bought up by mergers and acquisitions. But the unusually steep recent decline is due to the steady drop in the number of companies issuing stock through initial public offerings (IPOs), presumably due to a tepid response. Today's investors have a reduced appetite to research and seek out individual companies on which to risk their capital. Instead they are buying baskets of stocks via mutual funds and Exchange Traded Funds (ETFs).

Jack Bogle is credited with introducing the notion that turned the tables: If no manager can beat a benchmark consistently over time, why try? Instead of paying high fees for active management to try to achieve the impossible, he suggested, why not create a fund that would simply mimic an index such as the Dow Jones Industrial Average (DJIA) or the S&P 500 and achieve market returns with lower costs. "The Father of Modern Economics," Paul Samuelson, ranked Bogle's "invention" as a transformative notion as significant as the invention of the wheel, the alphabet, and the Gutenberg printing press.

Are You as Diversified as You Believe You Are?

Funds have grown explosively. As of mid-2019, altogether there were 10,572 mutual funds, closed-end, and ETFs available, most of which are essentially some combination of the 4,397 publicly listed companies. In May 2019, history was made when the amount of money in passive index funds matched the amount that was in actively managed funds.

Financial news stories often lead by reporting whether the DJIA is up or down, and most people look at "the Dow" as a proxy for how the market is faring. But although it is taken as a gauge of the health of all of corporate America, the Dow is a group of just 30 companies, and since its inception in 1896, the list of companies has been changed 54 times. Sears was removed in 1999 after a 75-year run, and in recent years, Kraft Food was dropped and replaced by United Health in 2012. In 2013, Alcoa, Bank of America Corporation, and Hewlett-Packard were replaced by Goldman Sachs, Nike, and Visa. Apple replaced AT&T in February 2018, and in June of that year, General Electric Company was replaced by Walgreens Boots Alliance.

The Dow is a price-weighted index, so companies with the highest-priced stocks have a much bigger influence on how the index moves. A $200 stock that goes down 20 points has far more effect on "how the Dow is doing" than a $20 stock that goes down $2, even though both companies have lost one-tenth of their value. An inexperienced investor might not realize that by buying "the Dow," you're benchmarking your own portfolio to just 30 companies that have very little relevance to the whole economy and especially not to your particular situation.

Because the DJIA is an imperfect representation of how the overall market is doing, the overwhelming majority of people looking for diversity turn to the S&P 500, and most invest in one of the three most popular ETFs—SPDR, iShares, and Vanguard—that are offered respectively by the giants of the financial industry, State Street, Black Rock, and Vanguard.

Many see "500" in a fund name and think that by investing in that fund, they're diversified—in other words, they think that their investment is spread among hundreds of companies. But the S&P 500 Index has its own pitfalls because it is cap-weighted, meaning that the total market value of a company (the number of shares outstanding multiplied by the share price) determines the relative size that company represents in the Index. As a result, 51% of the Index, or 51 cents of every dollar you invested in one of the aforementioned S&P 500 ETFs, goes toward the purchase of just 50 companies out of the entire 500 as of late 2019.

Even more significantly, 22% (22 cents) of that 51 cents goes to the purchase of just ten companies. And the larger a company gets, the larger a share of the fund goes to buying that stock. In the last few years, the S&P 500 funds have become very heavily concentrated in relatively few securities, notably Facebook, Amazon, Netflix, and Google, referred to as the FANG stocks—again, not a true picture of the market overall. Investors trying to buy or sell based on their changing perceptions of risk may be unaware that what they are actually doing is buying the winners in ever-greater quantities and they are not nearly as diversified as they had thought.

Aside from the DJIA or S&P 500 index funds, investors can choose among thousands of other index funds, each of which is designed to follow certain preset rules that dictate the purchase or sale of a specified basket of underlying investments. A fund might consist of stocks in the same industry or product line that are similar in size, value, or geographic location or some other characteristic. The issue here is that ETFs are traded like stocks, so liquidity (the ease of buying and selling the ETF) is essential. This means that although a fund has a name that suggests it is focused on environmentally responsible companies, it may own a significant amount of Exxon and/or other more liquid securities that do somehow fit into the rules set by the fund. Fund investors who don't check the fund's holdings may not realize that it owns securities that not only are unaligned with their philosophy but also duplicate the holdings they have in other fund baskets.

Here's the real danger: Pooled investments such as mutual funds or ETFs are an efficient way to own many securities, but they are also an effective way to create herds. In January 2017, a *Financial*

Times article entitled "ETFs Are Eating the U.S. Stock Market" reported that "ETFs now account for about 30% of all US trading by value, and 23% by share volume."[159] And that trend has only gained momentum since then. Overconcentration risk isn't an issue when markets are doing well. But when there's a correction it's a huge issue, because so many people may be trying to unload funds containing the very same stocks at the same time. On February 5, 2018, in one 20-minute period, everyone was heading for the exits. The market fell 1,600 points—and then rebounded 800 points by the end of the day. Such a colossal, rapid swing could not have happened in the days before computerized trading.

Besides, if beating the indexes is your goal, you are easily misled. In the course of the long bull market run that began in 2009, there were multiple corrections in which many individual stocks and sectors had sharp declines. It is likely you overlooked downward trends in certain areas and companies with growth potential in others. By mid-2019, you might not have realized you were benchmarking primarily to huge tech companies that have relatively weak or no earnings. That's not the right concentration for retirement funds. To protect and grow your nest egg, you should always be thinking about which individual companies you want to put your faith in long-term.

Asset manager Marshall Jaffe summarized the situation perfectly: "Markets are efficient pricing mechanisms because the majority of participants are making their choices independently. Diversity of input is what creates a 'smart crowd.' When that diversity breaks down and too many participants make their choices from a limited number of shared sources, the smart crowd starts behaving like a mob and the pricing mechanism loses its efficiency."[160]

"If everybody indexed, the only word you could use is chaos, catastrophe. The markets would fail." You know who said that? John Bogle—yup, the guy who *invented* indexing.[161]

For Better and For Worse, Technology Speeds Up Investing

As I mentioned, when I bought my first stock, I had to place my order through a middleman, my father's broker. When I entered the financial field a few years later, buying stock was still so labor-intensive that there could be a significant lag between your initial contact with a broker and the execution of your order. Brokers were obliged to fill out a buy ticket that had five copies, each color-coded for a different department, before they could even send your order into the system.

Recently, a friend's 19-year-old son told me about his own first stock purchase: He linked a stock trading account to his savings account—holding just $300—and immediately made a transaction via smartphone. The introduction of technology accelerated the pace of buying and selling exponentially.

In 1968, when a then-record-breaking 13 million shares were traded in a single day, the resources of the NYSE were overwhelmed. After that so-called "Paperwork Crisis," for several months the Exchange closed down every Wednesday and also occasionally closed early on other days to deal with the backlogs. Still, because of so many "fails"—failures to receive or deliver securities within five days of a trade—one-sixth of firms were obliged to close or to merge. Today, computers and centralized electronic stock delivery handle NYSE trading volume of over a billion shares a day and NASDAQ average daily volume of two billion shares (and, on busy days, over five billion).[162]

In retrospect, the old ways of buying and selling and monitoring stock prices, which today seem tremendously inefficient, had a couple of significant benefits. Having to go through a broker to place an order gave you a chance to get counsel from a gatekeeper with experience. In addition, that broker's license obliged the broker to consider the interests of the investor, which was very helpful, especially when the investor was inexperienced.

For instance, if I were suggesting a stock to a client, or the client wanted more information about an investment she was considering, I'd send along an S&P tear sheet or a ValueLine report. It would take a few days for it to arrive in her mail, and then we'd discuss the merits of the investment and decide what to do. It was routine for people to do their homework before investing their money, and because they did, they really understood where they were putting their money and why, and they were committed to the choices they had made.

Also, by its nature the old-fashioned, laborious, and paper-intensive process of buying and selling stocks put some time between when you made your decision and when it was implemented, which built the possibility of deliberation into the process.

The steadfastness, discipline, and motivation of the parents of the baby boomers helped them become such successful investors. Today, investors by and large have so little connection with any of the securities they own that they tend to behave as if they're in a casino: all in or all out. They are likely to respond exactly the same way to any sort of disruptive event, whether it's a political upheaval, an industry slowdown, a hurricane, or a terrorist incident, although each could have a completely different effect (or even *no* effect at all) on an individual company.

Although technology cut the costs of middlemen, made the tools of investing available, and simplified the mechanisms, it reduced the deliberation and insight with which investors formerly acted. Today they often dumb down their actions to one-click solutions. That's unfortunate. The purpose of technology is not to help you to ask *no* questions but to help you to ask *better* questions. What is important in investing isn't having all the answers. It's asking the right questions to get the answers that are right for your needs and your situation.

It is easy to dismiss my thoughts on computerized passive investing as coming from a guy who makes his living charging for advice. But in fact, there is actually no such thing as "passive" investing. Whether you put your money into individual stocks, a fund, bonds, Bitcoin, or even a savings account, each choice represents an investment decision. You should know which is appropriate to use and when, and you have to understand the structure and possible consequences of the investments you make, especially in these volatile times.

To Protect Your Tomorrow, Ignore the Noise Today

Once I had bought my first shares of stock, naturally I was anxious to follow "my" stock's activity. I could check the *Wall Street Journal* every morning, of course, but that only reported the previous day's close. To get more frequent updates, I had to bother my dad's broker several times a day.

By the time I myself had become a broker, real-time quotes became available, though to get the very latest, you'd have to hit the "refresh" button on your Quotron machine. Today, of course, paper copies have long been obsolete, and transactions are made in nanoseconds; it is unthinkable to make a trade based on a quote that is even 10 seconds old. Investors today can even track the movement of their holdings in real time—on TV, via computer, smartphone or smart watch.

However, there's a big downside to the steady stream of information and commentary available, not just financial stories but the totality of the news that comes at us, every day and all day long, and compels so much of our attention.

Not only has technology brought us an explosion of media platforms—hundreds of them, each delivering messages from multiple sources—but it has also brought us streaming social media, that huge and astounding hodge-podge of news and fake news, rumors and opinions, all with the potential to excite emotions from jubilation to despair.

In years past, when you read the newspaper in the morning and found out what had been going on the day before, it was too late to act impulsively. But along with the noise, technology has brought us the means to react to it as quickly as our fingers can operate the keys on our devices.

On December 27, 2018, shortly after 9 p.m., my daughter called in a panic from her apartment on Manhattan's Upper East Side. The sky outside her window was as bright as if it were mid-afternoon, glowing a weird greenish-blue. She texted a photo and said that there were multiple similar images on Twitter, along with reports of a 9/11-type attack, a plane crash at LaGuardia, and Homeland Security halting all traffic to and from Queens.

I immediately logged onto my laptop, hoping to get confirmation from the traditional sources for breaking news: the major TV networks, local radio stations, and newspaper websites. There was no mention anywhere of what my daughter was witnessing. It was well over an hour later when these sources gathered the earlier facts and reported that a fire had caused an explosion at a Con Edison plant in Astoria, with no injuries or apparent foul play.

This was a perfect illustration of the transformative impact of the technology revolution. Now, any one of the four billion cell phone owners can be a broadcaster, within seconds spreading news to a network of friends who in turn reach out to their networks, and so on.

Because the explosion came long after the 4:00 p.m. close of the markets and long before it opened at 9:30 a.m., there was plenty of time to ascertain that what happened was of no serious consequence. Imagine, though, the dramatic and immediate negative effects if reports of another possible terrorist attack on New York had come when the markets were open and trading.

We are just beginning to understand how destabilizing the proliferation of raw, in-the-moment reporting of "facts" that may be inaccurate, incomplete or just plain wrong can be on financial markets. Unwarranted panic about a nonexistent "terrorist attack" is just the sort of thing that can cause the gut-churning ups and downs, sometimes hundreds of points in an instant, that have recently become commonplace in the financial markets based on rumors, tweets, or a simple phrase or two.

The Fed says, "We will be patient," or the president drops the phrase "trade war," and the roller coaster takes off. Thanks again to technology—which broadcasts the news and digitizes it almost simultaneously—the algorithms have already responded to it by the time you see the "Breaking News" banner scroll across the screen or get a push notification on your smartphone. You're virtually guaranteed to be in the back of the pack running for the exits (or, on some occasions, to the entrances).

The delay might be to your advantage, because it's too late to act on your emotions. Quick action is not for you. It's for the pros in the Wall Street sector trying to move fast enough to turn pennies on a trade into huge profits, a fat bonus, and/or a reputation as a benchmark beater. Your goal is different. Nothing about nest egg building happens in the short term. In fact, immediate reactions are usually the enemy of wealth creation.

Interestingly, investors are often able to tune out current noise when dealing with their 401(k)s because those funds can't be touched for a long time. They're out of sight, thus out of mind. But with investments that are not tax-deferred, people are more likely to try to time the markets by making frequent trades—despite the futility of such attempts.

In the past, the structure of the system made people more likely to stay with their investments. If they had had the ability to check on their investments as frequently as we do today, they might not have held so tightly to their shares of IBM and GM (which eventually paid off so remarkably well. And as a side benefit of holding on, they saw their gains taxed at favorable long-term rates).

According to the NYSE Factbook, the average holding period for stocks in 1960 was 100 months (8 years). By 1970 it had dropped to 63 months (5 years). By 1980 it had dropped to 33 months, by 1990 to 26 months, and by 2000 to just 14 months. The number has gone lower since. In 2016, the World Bank said it was only 8 months, although Credit Suisse estimated it was only 4 months.

Neither feeling challenged to try and outsmart a market driven by algorithms nor so discouraged by it that they moved to the sidelines, previous generations prospered. Success as an investor requires enough optimism to take the long view, use time to your advantage, and buck the tide to buy when others are selling. Messages that kill your motivation also kill your ability to make money. It's hard to keep your vision on the long-term when the short-term seems fraught with danger, when you're reacting to tweets, and when panic is driving your decision about when and whether to get out or get back in.

Hindsight reveals that often the best way to make money on an investment is not by taking action but by doing nothing. But technology is about action, and many investors who were able to identify a promising company often sell the stock long before it reaches its potential. You're always only a few clicks from going nuclear with your current assets because of a powerful instinct in our genes.

Loss Aversion Is Baked Into Our DNA

In the course of evolution, humankind spent eons living in unpredictable circumstances, and as a result, scientists tell us, we are hardwired to be far more uncomfortable at the prospect of facing a loss than we are inclined to take a risk for a possible gain.

A front page of *The Wall Street Journal* that appeared in August 2015 had a six-column headline: "Markets Reel in Global Selloff" followed by several more pages of analysis and related coverage. In three days, the market had lost 1,300 points from the cumulative effects of a variety of factors weighing on the minds of investors at the time: slowing growth in the GDP of China, a fall in petroleum prices, the Greek debt default in June 2015, and the effects of the end of qualitative easing in the United States in October 2014.[163]

Within five days, *The Wall Street Journal* ran another front-page story that delivered a very different message. It began, "The big question worrying investors now is whether last week's rally is sustainable or just the prelude to another storm."

When the initial headline appeared, I am sure that plenty of investors fearful of getting wiped out considered liquidating their holdings. No doubt many did. The consequence of loss aversion is panic selling, and the consequence of emotional selling may be major damage to your long-term investment goals: During a momentary steep decline you may lock in exaggerated losses. If you panic and take a 50% loss, the stock has to go up 100% for you to just break even.

Dalbar, a company that has been producing its highly respected *Quantitative Analysis of Investor Behavior* for a number of years, recently focused on the topic of investor psychology: "the psychological traps, triggers and misconceptions that cause investors to act irrationally."[164]

Investors in general are prone to a variety of counterproductive behaviors, though their unique experiences and personalities determine which will affect them and to what degree.

I've already discussed herding and loss aversion, two of the most common behaviors that Dalbar identified. And I have mentioned insufficient diversification: thinking you've spread risk among different kinds of investments when your assets are in a pool with a lot of duplication, which often is the case when you're investing in ETFs. But there are many others:

- *Narrow Framing*: Making decisions about one part of the portfolio without considering how those decisions will affect another part.
- *Mental Accounting*: Taking excessive risk in one area and avoiding rational risk in another. For example, people often decide to sell a winner (which may have additional potential) because they've already made a profit,

yet they'll hang onto a loser (with poor prospects) rather than take a loss.

- *Anchoring*: Clinging to behavior that worked in the past without adapting to changes in the market.
- *Regret*: Hesitating to make a necessary move due to remorse about having taken a previous action that failed.
- *Media Reactiveness*: Reacting reflexively to news without processing what the real consequences are to the market and your own holdings.
- *Excessive Optimism*: Believing that good things happen to me and bad things happen to others.

The one that has been paramount in the last decade is recency bias—using recent results to make incorrect conclusions and thus incorrect decisions about how the market will continue to behave.

As a result of the Great Recession, the percentage of Americans owning stock went from the 2007 high of 65% that I referred to a few pages ago all the way down to 52%. And for some people, the ghost of that event lingered for quite a while. They stayed on the sidelines with their cash, and many millennials, in particular, remained reluctant to invest because they had seen so much negativity. As a result, they missed out on one of the biggest bull runs in history.

Recency bias can operate in both directions. As a result of the market's decade-long upward climb, by 2019 many investors—convinced that they were Warren Buffets in the making—became less cautious and concluded that investing was easy. As the saying goes: "Don't confuse brains with a bull market." While over time the stock market has climbed steadily, it has always done so in a sawtooth pattern, with upward moves followed by a correction, and on the heels of that, a recovery that surpassed the prior high. It will be interesting to see how the whole generation that came to Wall Street in 2008 deals with the inevitable shakeout.

Managing Your Emotions Is the Key To Managing Your Money

The stocks or funds you pick are not the most important factor in your investment success, Many studies report, and my personal experience in the financial industry confirms, that people make their investment decisions in a more visceral way—how they feel in their gut, based on the atmosphere of the moment. If there's a sudden dip in the market, a trade agreement that doesn't seem to be working, or a government shutdown, they're bearish. Conversely, after an extended bull market, they are inspired by the general euphoria to get back in, very likely at a point when prices have already risen dramatically.

People routinely react this way even though it directly contradicts the most fundamental law of investing: Buy low, sell high. Selling at the wrong time invariably leads to underperformance. What I said a page or two ago is worth repeating, to drive the point home: A stock that goes down 50% has to go up 100% just to get back to the starting point.

What makes people susceptible to emotions is one basic thing: They don't know who they are as an investor. And that's more likely than ever to be true today, when many people seek financial advice online.

The Perils of Online Planning

Financial management firms have begun to use technology in financial planning in much the same way as they have used technology to change the nature of investing. Similar to the way in which they create baskets of funds rather than looking at the merits of individual companies to make a portfolio, the firms have created "baskets" of investors. They collect data from prospective clients (typically via an online questionnaire), group them, and offer a generic plan to each group. It's very efficient for them, but it has many of the following pitfalls for an investor:

- *Much depends on your interpretation of the questions.* To proactively control what a robo-advisor does with your money, you have to be reflective and truthful when it comes to filling out the questionnaires. But if you don't really know yourself as an investor, you can be sure the robot analyzing your responses won't know you either.
- *Advice may be based on incomplete information.* One way to gut check the value of the website is to consider the information the online questionnaire is soliciting from you. If it's *not* the kind of specific information you'd probably feel uncomfortable about revealing to an unfamiliar source—your total income, the size of your bank account, what you're spending on credit cards and so on—there's no way the site has collected enough information to recommend a meaningful course of action for you.
- *Your answers to a questionnaire aren't a real reflection of your risk tolerance.* The notion is that online investing removes emotion from decision-making to eliminate bad and costly choices. However, the questionnaires are really designed to determine how much volatility you can withstand. And aspirational questions on the form (such as "What are your dreams?") are really marketing questions in disguise, not a basis for making a plan. If you reveal that your dreams include high-priced acquisitions, like yachts and diamonds, you're categorized as someone willing to take on a lot of risk. These questions are less about wanting to know about you and more about gauging the amount of pain you

can take from a failing portfolio before you are likely to sue.

In any case, people aren't sure about their appetite for risk. They tend to be overconfident when the market is good and fearful when it drops. Although it may sound counterintuitive, I suggest you answer the questions emotionally, rather than logically. For example, if the market goes down, you might understand it's a buying opportunity, but if the thought of the market going down gets your adrenaline pumping (and not in a good way), that's something to address. Asked what they can afford to lose, most people will say 5% or 10%, fewer will say 15%, and no one says 25%. One good way to gauge your risk tolerance is to consider your reaction to the Great Recession of 2008.

True, some questionnaires do ask questions geared to get more specific about your willingness to take on risk. They won't ask if you can afford to lose X amount of dollars, of course. Instead, they'll ask questions about human dynamics: "Does the market volatility keep you up at night?" "Do you feel the need to check the value of your portfolio daily?" Translation: Are you very risk-averse or not?

Basing your actions on online recommendations is appropriate when the consequences are short-term—such as one inedible restaurant dinner, a week at a disappointing hotel, even a couple of years driving a car that fails to meet expectations. But bad financial advice could crush your whole life.

Long-term planning should help you respond to ever-changing circumstances. The future is not always what we expect. And like Mr. Spock, algorithms are clever and analytical and superb in matters of logic and calculations, but they lack nuance. Computers are very good at answering specific questions. They can tell you the best route to get somewhere, but they can't tell you if going there is a good idea. They can calculate that you can afford $500,000 in life insurance or a $700,000 house, or that you shouldn't buy a boat longer than 80 feet. But they won't tell you that since you have no children, you don't need such a big policy, suggest that you stay in your paid-off home rather than take on more debt, or point out that you haven't got any emergency funds put aside.

It's fairly simple to predict how much money you need to live on twenty years from now by estimating budget items, factoring in inflation, and then calculating an appropriate rate of decumulation. It's quite another to contemplate the appropriate way to manage your money in response to some unpredictable event—a failing business venture, a debilitating illness, a costly divorce, a child overwhelmed by student debt, or another unfortunate real-life situation.

That is why the questionnaire approach can never give you advice that is truly suitable for you. After all, the number of people whose answers were identical to yours would most probably fill a space even bigger than Yankee Stadium (perhaps even multiple stadiums). Yet no two of you are likely to have the same needs and goals.

You Must Be a Participant in Your Own Planning

Before you choose a planner, before you talk about risk tolerance, before you start thinking about companies and investment choices, you need to spend some time contemplating a picture of the future and coming to some decisions.

Ask yourself when you will need money: What can you earmark for the long term? What will you need for the short term? (If you need money short-term, you shouldn't be in the market.) How much should you set aside for emergencies? What are your needs? What are your priorities? What long-term goals are realistic? Until you have really given thought to these questions and have arrived at some answers, you can't put together a portfolio that you're motivated to stick with and possibly not even able to.

You need to learn who you are as an investor and build on that confidence by educating yourself and doing your homework.

When you have a real plan, you won't be shaken out when there's a setback.

I always say that the number one service I provide is not money management but keeping my clients focused on the long run, not distracted by the emotions of the moment. Emotions kill more wealth than bear markets.

Opportunities for investors have become greater, thanks to technological innovations that make it easier to trade stocks, commodities, and currencies domestically and overseas. But investment risks have become greater as well. Individuals notoriously have terrible long-term performance, mostly because they allow impulsive decisions to rob them of their gains.

A licensed advisor, in contrast, will do a deep dive into your finances, has fiduciary responsibility to act in your best interests, and is anticipating a working relationship with you. And the calm counsel of an experienced professional, someone who's lived through the ups and downs of the past, is likely to help you stay focused on your long-term goal: preserving your lifestyle for your lifetime.

For that goal to become a reality in this time of rapid change, it's important to think about the issues I have raised. True financial success and mental tranquility come from feeling positive about how you spend your time and resources. They come from knowing that your thinking and behavior is not led by others but by a consideration of the facts. They are achieved by being clear about your purpose so you act with deliberation rather than based on the mood of the moment and by educating yourself to be prepared for the world as it is becoming rather than the world as it was.

Postscript

Note: I have no doubt that as you are were reading this book you came across some information (statistics, references to websites, and so on)—and even opinions and suggestions that are already a bit behind the times. I know this because the rapid pace of technological change necessitated my updating this manuscript on a regular basis, up until the day we went to press. But I believe that the concerns and advice will remain relevant.

WHETHER A CHANGE IS LASTING AND POSITIVE is not always immediately apparent. History is full of great new developments that with the passage of time turned out to have some unanticipated downsides.

For example: In the decades following the Industrial Revolution, towns across America welcomed and even incentivized factories to locate nearby, welcoming the economic shot in the arm they gave the community. Only after many years did it come to light that the waste products that the factories dumped had polluted the local soil and water, causing devastating diseases and even birth defects.

X-rays, invented in 1895, were first introduced to the public as a carnival novelty and then adopted by the shoe industry as a sales gimmick in the 1920s. Customers would try a on pair of shoes and slip their feet into an X-ray machine. Then salesmen would look into the scope to assess the fit. After scientific reports emerged reporting the health consequences of radiation exposure, the machines were banned.[165]

In the 1930s, Phillip Morris advertisements alleged that after doctors prescribed its brand to people with irritated throats, "every case of irritation cleared completely or definitely improved." Camel boasted that "More doctors smoke Camels than any other cigarette" and placed ads recommending the smokes as a digestive aid. Then, in 1954, two scientists working for the American Cancer Society published a study definitively establishing a link between cancer and smoking. From a peak of 42.4% of adults who smoked in 1965, the number of smokers has decreased steadily, down to 14% when the CDC issued its latest report in November 2018.

Plastic bags were introduced in the 1970s as a cheap, "green" (because they saved trees) alternative to paper bags. A half-century later, 500 million of them are used annually, most discarded as litter, either sitting in landfills and not likely to decompose for hundreds of years or befouling our oceans and threatening uncountable species of marine life.

Text messaging is among the latest innovations to pose unexpected hazards. Introduced in the 1990's for quick communication, it has become an indispensable everyday tool. But it is estimated that 25% of car accidents in the United States are caused by texting while driving, which by some estimates is six times more likely to cause a car accident than drunk driving.

We're just becoming aware of some of the unintended consequences of the technological transformations I've mentioned. We are addicted to our screens, obsessed with social media, and disconnected from how we spend money.

We are recognizing the threats that result from the ubiquity and influence of technology in our lives. Originally, we were cavalier about giving up privacy to unknown entities as we clicked on our keyboards—"Who cares if 'they' collect data on what I buy or where I want to go on vacation?"—but we have a growing sense that everything we do gives off digital exhaust that can be tracked and used to manipulate us in significant, sometimes ominous ways. And the manner in which "news" is collected and the speed at which it is delivered to our increasingly connected world require us to be more cautious about what we allow to influence our thinking.

We are just becoming aware of the tactics we need to adapt to a world of unceasing updates and disruptive changes. We must be educated—and reeducated—on an ongoing basis. We must be the custodians not only of ourselves but of our cyberselves. And technology requires us to find new approaches to how we finance, save, and invest for the future.

In my practice, I see daily how people are managing their money with the aid of technology, using tools that are available to anyone regardless of their understanding of how best to use them. This concerns me.

It is tremendously positive that more people have this opportunity, but in this regard, as with all the others I have raised in the book, it will be years before we learn of the unintended consequences of the choices we make. I wanted to highlight some of the ways the pace of technological change is simultaneously enhancing our lives and potentially threatening the future lifestyle we envision for ourselves.

We are lucky to be alive during the most exciting period of human history. The development of the computer chip has bestowed upon us abilities and benefits our ancestors could never have imagined. And we of course can't predict exactly how artificial intelligence, virtual reality, machine learning and other developments will impact society over time.

Which brings me back to my observations about those migrating Canada geese at the start of this book. They have evolved to function brilliantly, but they are creatures of instinct. We humans, alone among species, are able to think about the future, reflect on change as it happens, and take actions to adapt to it both as a society and individually.

Unlike the geese that need to wait millennia for the passive forces of evolution to help them deal with the perils they face, you can take control and travel safely and thoughtfully on the digital highway, avoiding forces that can crack your nest egg.

Sources

Introduction

1 https://www.newscientist.com/article/dn9989-timeline-human-evolution/
2 https://en.wikipedia.org/wiki/Native_American_use_of_fire
3 http://www.ancient-origins.net/ancient-technology/revolutionary-invention-wheel-001713
4 http://www.ancient-origins.net/ancient-technology/revolutionary-invention-wheel-001713
5 *I'll Mature When I'm Dead*, "Technology," by Dave Barry, G.P. Putnam's Sons; Reprint edition (April 5, 2011)
6 https://www.recode.net/2017/6/26/15821652/iphone-apple-10-year-anniversary-launch-mobile-stats-smart-phone-steve-jobs
7 https://wearesocial.com/blog/2019/04/the-state-of-digital-in-april-2019-all-the-numbers-you-need-to-know
8 http://www.wired.co.uk/article/wired-explains-moores-law
9 https://www.nytimes.com/2017/05/19/opinion/sunday/why-the-future-is-always-on-your-mind.html
10 https://www.nytimes.com/2017/05/19/opinion/sunday/why-the-future-is-always-on-your-mind.html?_r=0
11 http://u.demog.berkeley.edu/~andrew/1918/figure2.html)]
12 https://www.teacherpensions.org/resource/insufficient-how-state-pension-plans-leave-teachers-inadequate-retirement-savings
13 https://money.cnn.com/retirement/guide/pensions_basics.moneymag/index7.htm
14 businessinsider.com/social-security-life-table-charts-2014-3
15 http://science.sciencemag.org/content/345/6192/75
16 www.bbc.com/capital/story/20170414-why-idle-moments-are-crucial-for-creativity
17 https://aeon.co/ideas/let-the-soul-dangle-how-mind-wandering-spurs-creativity
18 http://time.com/4960778/computers-smarter-than-humans

Chapter One

19 https://www.nytimes.com/2017/07/03/upshot/why-some-men-dont-work-video-games-have-gotten-really-good.html
20 www.polygon.com/2016/4/29/11539102/gaming-stats-2016-esa-essential facts
21 https://www.drugabuse.gov/publications/teaching-packets/neurobiology-drug-addiction/section-ii-reward-pathway-addiction
22 https://www.npr.org/templates/transcript/transcript.php?storyId=557418328
23 https://lifehacker.com/change-your-screen-to-grayscale-to-combat-phone-addicti-1795821843
24 https://www.nytimes.com/2018/10/26/style/phones-children-silicon-valley.html
25 https://www.nytimes.com/2018/01/08/technology/aåpple-tech-children-jana-calstrs.html?rref=collection%2Fsectioncollection%2Fbusiness&action=click&contentCollection=business&region=rank&module=package&version=highlights&contentPlacement=6&pgtype=sectionfront&_r=0

26 https://www.commonsensemedia.org/sites/default/files/uploads/research/census_researchreport.pdf

27 https://www.pewinternet.org/2018/08/22/how-teens-and-parents-navigate-screen-time-and-device-distractions/

28 https://www.marketwatch.com/story/people-are-spending-most-of-their-waking-hours-staring-at-screens-2018-08-01

29 http://www.slate.com/articles/arts/culturebox/features/2013/daily_rituals/is_the_key_to_becoming_a_great_writer_having_a_day_job.html

30 https://www.inc.com/nick-skillicorn/the-10000-hour-rule-was-wrong-according-to-the-people-who-wrote-the-original-stu.htmlH

31 http://www.edudemic.com/the-4-negative-side-effects-of-technology/

32 https://hbr.org/2017/04/a-new-more-rigorous-study-confirms-the-more-you-use-facebook-the-worse-you-feel

33 http://press.careerbuilder.com/2019-03-05-Forty-%-of-Employers-Plan-to-Hire-Full-Time-Workers-This-Year-Forty-Seven-%-Recruiting-Part-Time-Workers

34 https://www.recode.net/2017/4/17/15334122/netflix-sleep-competitor-amazon-hbo

35 https://link.springer.com/article/10.3758/PBR.17.4.479

36 https://psychcentral.com/news/2014/10/13/for-some-multitasking-may-not-hinder-performance/76094.html

37 https://geneticliteracyproject.org/2016/07/25/another-modern-myth-shrinking-attention-spans/

38 https://jamanetwork.com/jouornals/jamanetworkopen/fullarticle/2598633

39 https://www.webmd.com/add-adhd/news/20181126/adhd-rising-in-the-us-but-why

40 www.webmd.com/sleep-disorders/features/power-down-better-sleep

41 https://www.wired.com/2015/03/artificial-light-may-be-unhealthy/

42 https://www.salary.com/articles/why-how-your-employees-are-wasting-time-at-work/

43 time.com/3858309/attention-spans-goldfish/

44 deloitte.com/us/en/pages/technology-media-and-telecommunications/articles/global-mobile-consumer-survey-us-edition.html

45 www.careerbuilder.com/share/aboutus/pressreleasesdetail.aspx?sd=6%2F9%2F2016&id=pr954&ed=12%2F31%2F2016

Chapter Two

46 www.theatlantic.com/business/archive/2016/04/why-dont-americans-save-money/478929/

47 https://www.psychologytoday.com/blog/ill-have-what-shes-having/201405/why-do-we-buy-luxury-brands-and-how-do-they-make-us-feel

48 https://www.sciencedirect.com/science/article/pii/S2405844017317279

49 http://blog.aicpa.org/2016/09/you-fancy-huh-one-in-four-americans-envious-on-social-media.html#sthash.PFesp3wN.dpbs

50 https://www.prnewswire.com/news-releases/new-study-shows-user-generated-content-tops-marketing-tactics-by-influencing-90-%-of-shoppers-purchasing-decisions-300475348.html

51 https://www.allianzlife.com/about/news-and-events/news-releases/Millennials-Generations-Ahead-study

52 www.retaildive.com/news/study-80-of-gen-z-purchases-influenced-by-social-media/447249

53 https://www.nytimes.com/2017/11/26/business/media/advertising-social-media.html?_r=0

54 http://fortune.com/2016/09/23/average-cable-tv-bill/

55 https://www.consumerreports.org/tv-service/cable-tv-fees/

56 https://www.statista.com/statistics/652009/united-states-most-common-smart-home-features/

57 https://www.iottechnews.com/news/2018/may/15/research-us-consumers-smart-home-device/

58 https://internetinternetofbusiness.com/idc-spending-iot-2018/

59 https://timeline.com/gm-invented-planned-obsolescence-cc19f207e842

60 http://www.bbc.com/future/story/20160612-heres-the-truth-about-the-planned-obsolescence-of-tech?ocid=ww.social.link.email

61 www.nytimes.com/1992/08/08/sports/sports-people

Chapter Three

62 https://www.statista.com/statistics/270728/market-volume-of-online-gaming-worldwide/

63 https://www.finextra.com/newsarticle/32856/swedens-central-bank-prepares-for cashless-future-with-e-krona

64 http://time.com/money/4226174/kill-100-dollar-bill-500-euro-phase-out/

65 http://newyork.cbslocal.com/2017/12/28/new-york-city-cash/

66 https://news.gallup.com/poll/193706/americans-foresee-death-cash-lifetime.aspx

67 https://www.pewresearch.org/fact-tank/2018/12/12/more-americans-are-making-no-weekly-purchases-with-cash/

68 https://www.youtube.com/watch?v=rBXJ9Razofw

69 https://www.nytimes.cm/2018/01/21/technology/inside-amazon-go-a-store-of-the-future.html

70 https://www.vice.com/en_ca/article/4w5vp9/why-your-mental-health-is-making-you-poor]

71 https://www.nerdwallet.com/blog/credit-cards/credit-cards-make-you-spend-more/

72 https://www.elitepersonalfinance.com/average-credit-card-debt/

73 https://www.thestreet.com/personal-finance/credit-cards/average-credit-card-debt-14863601

74 https://studentloanhero.com/student-loan-debt-statistics/

Chapter Four

75 http://blogs.ubc.ca/etec540sept10/2010/11/29/the-evolution-o-advertising-from-papyrus-to-youtube

76 https://table.skift.com/2018/10/24/how-custom-tipping-screens-impact-wages-at-counter-ser-restaurants/

77 https://www.nytimes.com/2019/06/17/business/media/miguela-virtual-influencer.html

78 https://influencermarketinghub.com/influencermarketing/

79 https://www.businessinsider.com/how-much-kim-kardashian-charges-for-instagram-endorsement-deals-2019-5

80 https://www.nytimes.com/2018/07/15/technology/online-stars-brands.html
81 https://www.technologyreview.com/s/535831/new-technologies-persuade-in-old-ways/
82 https:///www.retalidive/com/news/amazon-now-dominates-google-in-product-search/531822
83 https://www.mavrck.co/influencer-marketing-at-amazon-what-every-brand-can-learn-in-2018/
84 https://www.nytimes.com/2019/01/26/style/amazon-reviews-vine.html
85 https://www.businessinsider.com/amazon-vine-program-sellers-pay-for-reviews-2018-4
86 https://www.nytimes.com/2019/01/20/technology/amazon-ads-advertising.html?login=email&auth=login-email
87 https://www.nytimes.com/2018/07/20/business/how-consumers-can-resist-companies-market-power.html
88 https://www.snopes.com/fact-check/subliminal-advertising/
89 www.nytimes.com/2012/02/19/magazine/shopping-habits.html
90 https://www.theatlantic.com/technology/archive/2013/06/skinner-marketing-were-the-rats-and-facebook-likes-are-the-reward/276613/
91 https://www.sparefoot.com/self-storage/news/1432-self-storage-industry-statistics/
92 https://cbsnews.com/news/google-home-amazon-echo-patents-track-listen/
93 https://www.nytimes.com/2019/03/09/opinion/sunday/10Webb.html

Chapter Five

94 https://slate.com/technology/2018/04/are-you-really-facebooks-product-the-history-of-a-dangerous-idea.html
95 https://www.wsj.com/articles/you-give-apps-sensitive-personal-information-then-they-tell-facebook-11550851636?ns=prod/accounts-wsj
96 https://www.theregister.co.uk/2019/03/29/lipreading_smart_speakers_just_what_no one_always_wanted/
97 https://www.inquirer.com/business/harmelin-media-google-facebook-amazon-duopoly-cable-satellite-comcast-20190613.html.
98 https://www.pewinternet.org/2018/03/01/social-media-use-in-2018/
99 https://hbr.org/2016/04/targeted-ads-dont-just-make-you-more-likely-to-buy-they-can-change-how-you-think-about-yourself
100 http://lorrie.cranor.org/pubs/readingPolicyCost-authorDraft.pdf
101 https://www.pewresearch.org/fact-tank/2014/12/04/half-of-americans-dont-know-what-a-privacy-policy-is/
102 www.wsj.com/articles/the-data-colleges-collect-on-applicants-11548507602
103 https://www.nytimes.com/2018/06/26/magazine/letter-of-recommendation-timehop.html

Chapter Six

104 https://en.wikipedia.org/wiki/Clarke%27s_three_laws
105 https://www.vox.com/xpress/2014/12/9/7360709/internetinternget-learning-sharing
106 ttps://www.pnas.org/content/112/33/E4512
107 https://blog.hootsuite.com/facebook-algorithm/
108 https://www.thrillist.com/entertainment/nationa/how-net-netflix-recommendation-algorithm-works.

109 https://www.nytimes.com/2017/03/21/books/the-death-of-expertise-explores-how-ignorance-became-a-virtue.html

110 buzzfeednews.com/article/charliewarzel/the-terrifying-future-of-fake-news#.vu4korR

111 https://science.sciencemag.org/content/348/6239/1130.full

112 http://eprints.imtlucca.it/3020/

113 https://www.nytimes.com/2019/01/10/us/politics/facebook-fake-news-2016-election.html

114 https://mediapsychology101.com/category/psychology/media-psychology-psychology/

115 https://andrewchen.co/friends-versus-followers-twitters-elegant-design-for-grouping-contacts/

116 https://www.rollingstone.com/politics/politics-news/nancy-pelosi-doctored-video-facebook-twitter-rules-839115/

117 https://ourworldindata.org/democracy

118 http://worldpopulationreview.com/countries/literacy-rate-by-country/

119 https://www.google.com/search?q=world+population+in+2019&oq=world+population+in+2019&aqs=chrome..69i57.8129j0j4&sourceid=chrome&ie=UTF-8

120 https://www.worldbank.org/en/news/feature/2018/12/21/year-in-review-2018-in-14-charts

121 https://population.un.org/wpp/Publications/Files/WPP2019_Highlights.pdf-https://www.cdc.gov/nchs/data/nvsr/nvsr66/nvsr66_04.pdf

122 https://www.cdc.gov/nchs/data/nvsr/nvsr66/nvsr66_04.pdf

123 https://www.aafp.org/news/health-of-the-public/20181210lifeexpectdrop.html

124 https://www.bcbs.com/the-health-of-america/reports/major-depression-the-impact-overall-health

125 https://psmag.com/news/research-suggests-trumps-election-has-been-detrimental-to-many-americans-mental-health

Chapter Seven

126 https://www.nytimes.com/2017/12/11/style/robots-jobs-children.html

127 https://www.ted.com/talks/david_autor_why_are_there_still_so_many_jobs/transcript?language=enhttp

128 http://www.pewresearch.org/fact-tank/2018/08/07/for-most-us-workers-real-wages-have-barely-budged-for-decaes/

129 https://research.upjohn.org/cgi/viewcontent.cgi?referer=&httpsredir=1&article=1302&context=up_workingpapers

130 https://www.nytimes.com/2018/05/16/opinion/college-useful-cost-jobs.html

131 https://www.ssa.gov/policy/docs/research-summaries/education-earnings.html

132 https://cew.georgetown.edu/wp-content/uploads/Exec-Summary-web-B/pdf

133 https://www.ssa.gov/policy/docs/research-summaries/education-earnings.html

134 https://www.brookings.edu/research/digitalization-and-the-american-workforce/

135 http://burning-glass.com/content/uploads/Digital_Edge_report_2017_final.pdf

136 https://www.nytimes.com/2017/05/10/opinion/owning-your-own-future.html

137 https://www.futureislearning.com/media

138 https://www.classcentral.com/report/mooc-stats-2018/

139 https://hbr.org/2015/09/whos-benefiting-from-moocs-and-why

140 http://news.mit.edu/2017/mooc-study-offers-insights-into-online-learner-engagement-behavior-0112, https://hbr.org/2015/09/whos-benefiting-from-moocs-and-why.

141 https://www.classcentral.com/report/mooc-based-degrees-vs-traditional-online-degrees/

142 https://www.nytimes.com/2018/12/06/business/economy/artificial-intelligence-hiring.html

143 https://www.insidehighered.com/news/2018/12/14/survey-finds-increasing-interest-skills-based-hiring-online-credentials-and-prehire

144 https://www.nytimes.com/2018/12/06/business/economy/artificial-intelligence-hiring.html

145 https://www.cnn.com/2019/01/22/success/ageism-women-work/index.html

146 https://www.nytimes.com/2016/08/18/upshot/as-more-older-people-seek-work-they-are-put-into-old-person-jobs.html?mcubz=3&_r=0

147 https://www.cnn.com/2019/01/22/success/ageism-women-work/index.html

Chapter Eight

148 https://www.internetworldstats.com/stats.htm

149 https://blog.dashlane.com/infographic-online-overload-its-worse-than-you-thought

150 https://www.aarp.org/money/scams-fraud/info-2019/survey-identity-fraud-decline.html

151 https://www.creditcards.com/credit-card-news/credit-card-virtual-account-numbers.php

152 https://www.javelinstrategy.com/press-release/consumers-increasingly-shoulder-burden-sophisticated-fraud-schemes-according-2019

153 www.iii.org/fact-statistic/facts-statistics-identity-theft-and-cybercrime

154 https://www.businessinsider.com/deloitte-study-91-%-agree-terms-of-service without-reading-2017-11

155 https://tech.slashdot.org/story/16/07/13/2010231/tos-agreements-require-giving-up-first-born-and-users-gladly-consent

156 www.businessinsider.com/sneaky-terms-from-tech-companies-2013-11#

157 https://www.howtogeek.com/403769/what-happens-to-your-digital-property-when-a-company-goes-out-of-business/

Chapter Nine

158 https://data.worldbank.org/indicator/CM.MKT.LDOM.NO

159 https://www.ft.com/content/6dabad28-e19c-11e6-9645-c9357a75844a

160 https://www.thinkadvisor.com/2015/10/02/how-google-makes-investing-worse-3/

161 https://www.forbes.com/sites/greatspeculations/2018/09/19/are-we-headed-for-a-passive-index-meltdown/#4ecf2711413e

162 https://finance.zacks.com/define-volume-stock-market-6839.html

163 https://www.thinkadvisor.com/2015/10/02/how-google-makes-investing-worse-3/

164 https://www.dalbar.com/QAIB/Index

Postscript

165 https://www.healio.com/hematology-oncology/news/print/hemonc-today/%7B2df0dfa1-ce57-490a-8bea-6533f4a89f84%7D/the-shoe-fitting-fluoroscope-a-little-known-application-of-the-x-ray

Index

D

T

About the Author

photo © Kresimir Juraga

WITH NEARLY FOUR DECADES OF EXPERIENCE in the securities and personal finance industries, Ken Kamen is a nationally recognized financial expert and advisor. He is the author of the highly acclaimed book from Bloomberg Press, *Reclaim Your Nest Egg: Take Control of Your Financial Future.* He is president of Mercadien Asset Management as well as president of Mercadien Securities and was a past chairman of the National Investment Banking Association, the preeminent national association of regional banking firms.

Ken has consulted with the U.S. House of Representatives Commerce Committee and Ways and Means Committee, testified before the House Subcommittee on Finance, and served as chairman of a business advisory group to the U.S. House of Representatives' Oversight and Investigations Subcommittee on matters surrounding financial and accounting reforms as well as small business issues.

He is frequently called upon by national print and broadcast media to comment on a wide range of financial- and investment-related topics and trends. Ken appears regularly on CNBC, Reuters, and Fox Business Channel, has contributed to Forbes.com, and is often invited to colleges, universities, and many investment-related forums to speak on topics related to personal finance and global markets.